Flying Lightness

Flying Lightness

Adriaan Beukers
Ed van Hinte

Promises for structural elegance

010 Publishers, Rotterdam 2005

NEW YORK ROAD RUNNERS
1:45
SPEED LIMIT 45
ONLY

Fast futures

In the most prosperous and most densely populated areas of the world some people cover considerable distances in daily traffic, whereas for others work is just around the corner. So obviously the number of travelled kilometres varies. The acceptable travelling time, however, appears to be just about the same for everybody: an hour and a half per day. This is mainly due to the fact that we, like our ancestors (even the very ancient ones), travel and gather our income largely during the hours of daylight and work should not suffer because of the time spent on the road. The number of kilometres that a passenger travels per year increases with income. As a result, the average speed of travelling is rising along with it. When people cannot afford to exchange working hours for travelling time, or to combine the two for that matter, the increase in income entails a need for faster transport. This transition from low-speed to high-speed travel is becoming ever more apparent in Japan, Western Europe, North America and China. Initially slow trains and buses could hardly compete with the more flexible and faster car.

Sports brand Adidas claims, that 'Nothing is impossible'. They may in fact be right, for 92-year-old marathon runner Fauja Singh co-stars in a campaign for sneakers.

Now it turns out that the contribution of cars to passenger transport is coming to a halt and is even decreasing, at least in Europe and the US.

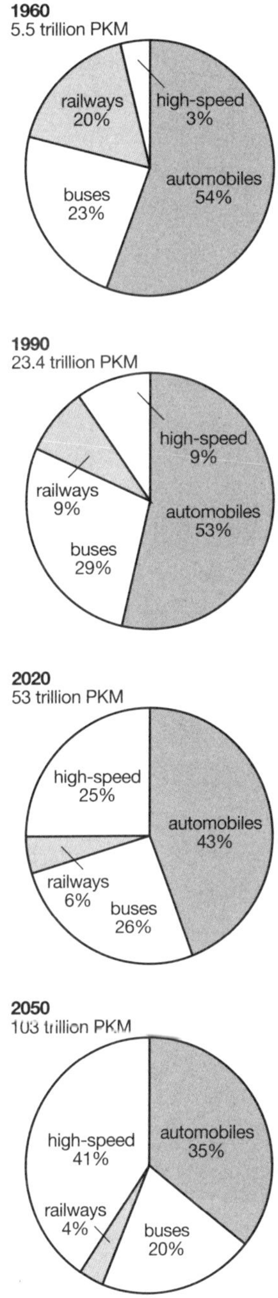

Over the past half century the share of high-speed passenger transport in a growing transport market has been continuously increasing.

The reason behind this is a silted-up regional and urban infrastructure, resulting in stagnating traffic circulation. As a result, the average travelling speed is decreasing and travelling is taking longer. This is all the logical result of the unrestrained growth of road transport, which is still very popular, relatively cheap and it takes you from door to door. The decline of efficiency can only be relieved by daring and sensitive decisions, such as congestion charge, pedestrian zones with transfer centres, and artificial intelligence in 'smart transport' with vehicles that 'know' their way around and have a way of avoiding obstacles. The transport sectors that do grow are those where the deployed measures result in time savings. Japan shows that a separation between the local and regional rail connections, as well as a separation between low-speed and high-speed transport, works very well. The Dutch government has always attempted to squeeze all transportation through one and the same infrastructure of regional highways and railways by maximizing the number of traffic hampering intersections. Fortunately the French and European TGV high speed train is increasingly asserting itself. This division of speeds is already customary in air transport, at no extra expense. In the future the ever busier continental connections will increasingly become the domain of high-speed transport by train and aircraft.

Traffic jams mainly occur where different travelling speeds meet.

On top of that, the international routes will remain the exclusive domain of the faster subsonic aircraft, but they won't look the same as the ones we are used to now, and possibly fly under entirely different conditions. This book is about the way in which they may become an everyday reality.

AN UNEXPECTED LACK OF CHAOS

Even busy roads need maintenance. For this reason the Dutch national government decided that most entrances and exits of the western part of the southern A10, the ring road around Amsterdam, were to be closed down to traffic completely during the summer of 2001. The behaviour of the authorities involved came close to initial panic. They expected gigantic traffic jams and launched a campaign to warn the drivers of the 180,000 vehicles that problems might occur, giving information about alternative transport and extra buses. What really happened at the end of May 2001 was the ultimate anticlimax. First there were no signs of congestion in the trial weekend. So the organizers figured that the real problem would start on the Monday after, but ... nothing. And the strange thing was that the special buses remained empty and that although there was some extra traffic inside the city, congestion simply didn't happen.
The authorities were absolutely flabbergasted and lost for words, let alone explanations. The Minister for Traffic, Tineke Netelenbos, claimed to have been fascinated and that what had happened was nothing short of a miracle. Her ministry immediately started an investigation, you know, the way ministries do. In the meantime Ms Netelenbos surmised that her advertising campaign had worked. But the main lesson, she said, was to keep local and regional transport apart. The theory already was that mixing two kinds of traffic causes jams and now it had been proved in practice. The minister said that from now on she would be more reticent in promising highway access to municipalities. Now the traffic policy is very slowly changing towards a division between kinds of traffic with different speeds. There are plans for more alternative routes rather than just adding extra lanes to existing highways. More improvement can be expected in the future, when cars become smart enough to stay close at high speeds, thus maximizing road capacity. The ideal solution is of course the racing circuit with no entrance but the one from the pit stop.

Development of large-scale projects shouldn't suffer from surreal expectations of short-term promises.

Subsonic flight will, however, never entirely replace other ways to move stuff or people from one place to another. There will always be differences in efficiency that have to be accounted for, depending on speed and the amount of goods that have to be transported. By using the property of 'specific resistance', the resistance per unit mass, we can tell that for regional transport buses and cars will remain attractive, provided that their velocity is kept constant and low, paradoxically resulting in a higher average speed than at present. Smart route planning, together with a civilized control of driving behaviour (below 100 km/hour and with less aberration from the average), can make road vehicles more efficient and provide shorter travelling times.

Table 1a. Transport systems, velocity domains and resistance per unit weight
Proven and future transport systems

velocity domain [km/h]	transport market/ system	fuel	dominant resistance / drag	specific drag D_{spec}
50 < V < 100	LOCAL smart buses, smart cars, smart trains	gas and electricity	wheels (rubber/asphalt) (steel/steel)	0.01 < D_{spec} < 0.02
50 < V < 250	LOCAL, REGIONAL CONTINENTAL human controlled cars and buses	liquid fuels	wheels and air friction (rubber/stone)	0.01 < D_{spec} < 0.20
125 < V < 300	REGIONAL, CONTINENTAL high velocity trains	electricity	air friction (steel/steel)	0.01 < D_{spec} < 0.05
300 < V < 900 relative speed	REGIONAL, CONTINENTAL INTERCONTINENTAL subsonic aircraft	liquid fuels hydrogen	air friction & induced drag	0.08 < D_{spec} < 0.05!

D_{spec} definition: see table 1b, page 12

22

The corresponding rolling resistance at less than 20 grams per kilogram moving mass is hardly relevant. Smarter transport, rendering the driver redundant if necessary, will both make our road network safer and more efficient. On top of that, most of the heavy passive safety measures in cars will become superfluous, which in its turn will decrease fuel con-

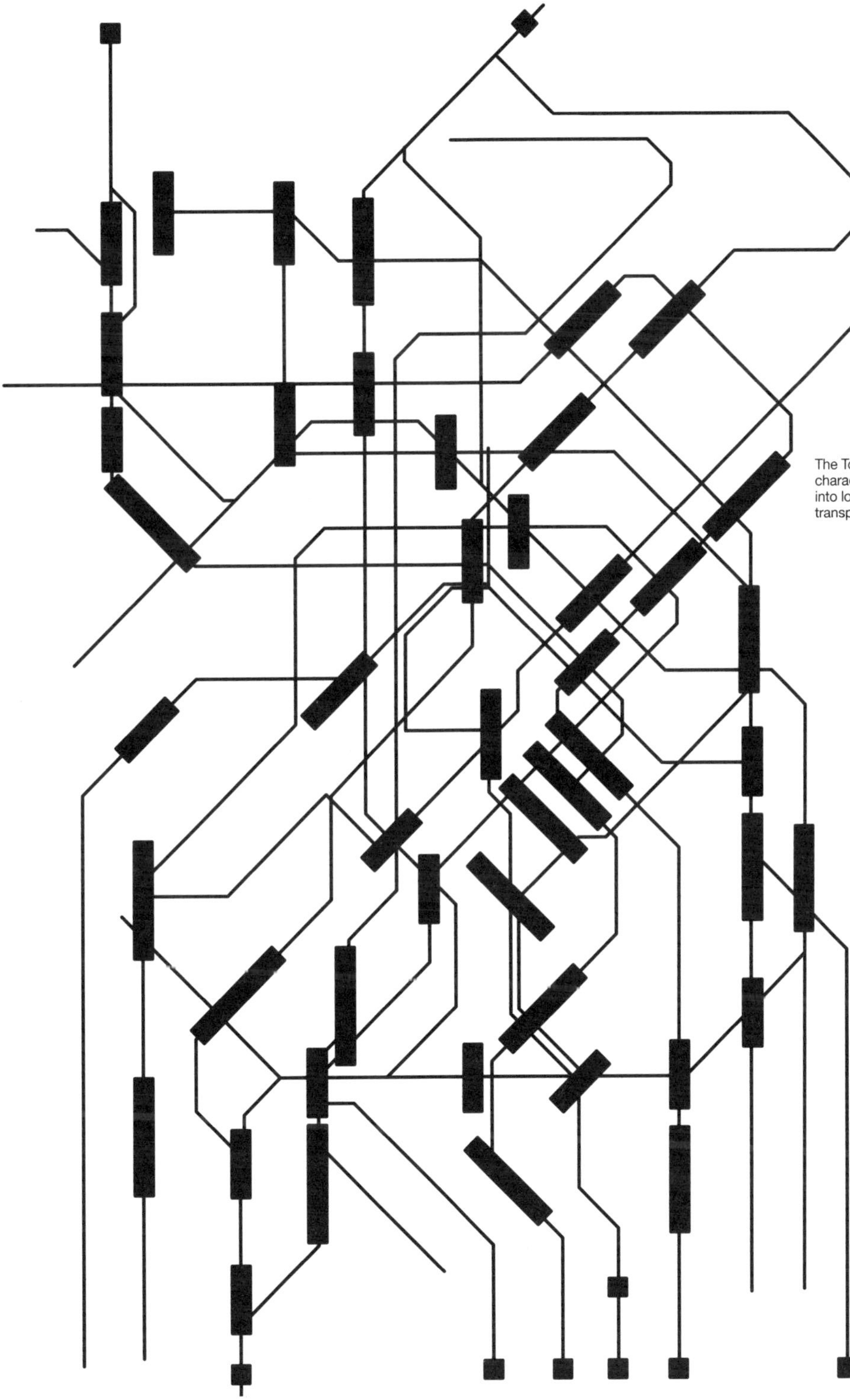

The Tokyo infrastructure is characterized by the separation into low and high velocity transport systems.

sumption and emission of polluting gases. According to the German car drivers association ADAC, a mass reduction of 200 kg results in a saving of about 1.5 litres of petrol for cars with a specific power of 1 W/kg at a speed of 100 km/hour. For regional and continental transport only a separate 'European' railway network will be able to provide the infrastructure for speeds of up to 300 km/hour. For this type of transport a maximum of 50 grams per kilogram moving weight is sufficient. If businessmen in massive Bentleys want to compete with this kind of efficiency, which is only possible on days when no other cars are allowed out on the streets, they will need more than 200 grams of every one of the many moving kilograms as a driving force. A pathetic figure, only exceeded at very low speeds by the very same small airships that are popular with Bentley driving businessmen and politicians alike.

NO AIRSHIPS

Because every moving vehicle has to overcome various forms of resistance, ranging from rolling resistance to drag from water or air, it is important to know how much of that has to be overcome per kilogram moving mass. By introducing a parameter called specific resistance (D_{spec}), the opportunity arises to classify different means of transport by the amount of the energy required to move one unit mass. Calculation is fairly simple: the force needed to overcome the resistance per unit mass, multiplied by the speed, provides the energy required to keep that unit moving.
By considering specific resistance we are able to classify vehicles in terms of energy efficiency and productivity (speed, passenger kilometres per hour). If the costs of energy consumption increase and their negative effects are taken into account, we obtain an indication of which means of transport will remain economically viable in the future.

Table 1b. Transport systems, velocity domains and resistance per unit weight
Exotic or obsolete transport systems

velocity domain [km/h]	transport market/ system	fuel	dominant resistance / drag	specific drag D_{spec}
V > 1,000 relative	INTERCONTINENTAL supersonic aircraft	liquid fuels	induced drag and wave drag	$0.10 < D_{spec} < 0.15$
V < 150 relative	REGIONAL wings in ground effect	liquid fuels	air friction and induced drag	$0.20 < D_{spec} < 0.30$
V < 120 relative	REGIONAL airships	liquid fuels	air friction	$0.025 < D_{spec} < 0.20$

D_{spec} = 0.1 means: for 1 kg (mass) in transport about 1 N (force) is needed to overcome resistance
smart could mean: speed and separation control, or satellite navigation control, or active instead of heavy passive safety precautions

22

For continental and intercontinental transport, and for the privileged that can afford to consider their travelling time too valuable, the only other option is subsonic flight. The current aircraft with turbo-fan engines, in which flying faster is linked to flying at higher altitudes with less drag (because of lower air density), the resistance as a fraction of the moving mass is at least 0.05. In the short term, new aircraft con-

The resistance D_{spec} as a fraction of the moving weight is represented in table 1 for several means of transport. It immediately shows that a number of them stand no chance of becoming economically viable because the D_{spec} is too high, for example more than 0.10 or 10%. Supersonic aeroplanes, ground effect aircraft, watercraft that skim over the water and small airships are doomed to remain exotic toys.

Obsolete lightness: at an auction of spare parts a Concorde nose cone fetched more than 500,000 Euros, nearly 10 times the expected price. Hungarian-born but US-based civil engineer Ferenc Gaspar became the proud owner.

A rich woman can afford
to travel in the equivalent of
4800 full handbags.
RR 1966
SILVER

DRIVING LITE

Apart from having minimum specific resistance, it is important that mass in motion consists primarily of load that really brings in money. The 'dead', non-productive structure requires exorbitant amounts of energy to move around for the duration of its economic life. In table 2 the ratio of the weight of the empty vehicle, W_{empty}, and the maximum amount of payload, $W_{payload}$, is classified for a number of means of transport. It demonstrates that the current bus, though without its passive safety measures, performs reasonably well at 2.5. In comparison, one man carrying a weight equal to his own will achieve something like 1.0, but with a very low productivity. He will be exhausted in minutes. By contrast an ant can carry a multiple of its own body weight resulting in a ratio of less than 0.5. The contemporary car is not very good at this game. Even for a full car 3.0 is far too much, not to mention the 12.0 for a car with your usual less than maximum payload: just the driver. Newly presented concept cars, like the VW 1 litre / 100 km, demonstrate with an admirable 2.0 what awaits the future economical driver. The ratio becomes even more favourable when an extra passenger is taken on board.
In the most expensive segment, which for cars rather oddly is also by far the heaviest, the ratios of 27.0 for the driver and still a solid 8.0 for three passengers can be considered poor rather than chic. In cars heaviness is still associated with high quality, durability and passive safety. Fortunately, most other consumer goods such as cameras, telephones, portable computers and sports equipment don't have these conservative connotations.

figurations could reduce this number by 20 per cent. If the empty weight is further reduced by the intensive application of fibre reinforced polymers, payload and efficiency can even be increased by some 30 per cent.
The fact that trains could be made lighter or exchange much of their dead weight for payload, in compliance with the vehicle dynamics, is increasingly appreciated by designers in this sector but is not always put into action. The term 'light-rail', popular among many politicians, is therefore merely a buzzword, as these trains weigh a despicable 1000 kg per passenger.

Table 2. Transport efficiency of different 'state of the art' transport systems

state of the art metal transport vehicles	W_{empty}/ $W_{payload}$ indicative#	specification
buses	2.5	
cars	3 (12)*	average car
	8 (27)*	Mercedes Benz S-class, 1st edition value dominated by propulsion and systems weight
subsonic aircraft	4"	balanced division of weight fractions#
supersonic aircraft	12"	value dominated by propulsion, systems and fuel weight
intercity trains	10"	value dominated by structural weight
global orbit	66	value dominated by fuel weight
lunar orbit	500	value dominated by fuel weight

* between the brackets the value is given for one occupant only
" values for 70% utilization
\# $W_{total} = W_{empty} / W_{payload}$
$W_{empty} = W_{propulsion/systems} + W_{fuel} + W_{structure}$

Although the present empty weight to payload ratio of 4.0 for aircraft already sets an example for heavier, slower vehicles, we can greatly top off this number by converting the classic metal tubular aircraft concept into a tailless flying wing, manufactured with fibre reinforced polymers. The potential to improve transport efficiency by as much as 20 to

40 per cent brings the standard of light city coaches within reach, but let us not forget that the productivity, the number of passenger-kilometres per hour is considerably higher. This change will take time, just as it took about 20 years for light metal aircraft structures to distinguish themselves from the original wooden 'flying crates' and contraptions of linen and steel tubes. In the same way the ideal composite aircraft will eventually look totally different from what we've become accustomed to. By then we can expect great improvements in performance, reduced operating costs and a smaller environmental footprint.

The weight of an articulated lorry just about equals its potential payload ($W_{empty}/W_{payload}$ equals 1), which turns it into an efficient transportation structure.

As far as the exploitation of aircraft is concerned, the trend that began in the 1960s of ever increasing costs and ever decreasing profits per seat, resulting in shrinking operating profits, must be radically altered. For most of the traditional hauliers time is running out and a further rationalizing of the management and optimization of the aircraft and air routes offers no solace. The new competitors know the same tricks and it is the pioneer entrepreneurs, with a long-term growth strategy in the back of their minds, who contend themselves with lower returns on the invested capital and use the net profits for reinvesting instead of 'banking' (the price fighters).

Whatever the appearance of air transport in the future, the companies that acquire and manage aircraft more efficiently will run the show, aided by the prescribed noise and emission reducing measures. Stricter and globally enforced regulations (for which again political courage is needed) would

CLEAN DREAM MACHINES

Air is crucial for planes to be able to fly. Unfortunately it is also a drag, a 'parasitic drag' to be more precise. That is what John K. ('Jack') Northrop fought against for most of his life. His passion was to build 'clean' aircraft as he called it. He grew up together with the airplane industry and was a keen technician from a very young age, interested both in physics and the practicalities of making things work. As soon as he had the chance he started working as a garage mechanic and later talked his way into working for an airplane company named Loughead. They built double-deckers, but already Jack Northrop put his efforts into eliminating protrusions that caused air resistance. The first design he got involved in was the F-1, a flying boat with a then novel veneer monocoque fuselage. He teamed up with 'Tailless Tony' Stadlman, and in 1923 they decided that they wanted to design the perfect wing, or rather the flying wing. For Loughead they made the S-1, a sports plane and the true first attempt towards a cleaner airplane.

Because of market changes both men took a job at Douglas and later Lockheed where they designed the Vega, which can be considered the second step towards cleanness. The first Vega was purchased by newspaper tycoon George Hearst, who helped Northrop to found the first company of his own, the Avion Corporation. There Northrop built the Avion Model 1, which already began to show traits of a flying wing but it still had to have a tail. It was an aluminium structure and had retractable landing gear which was later replaced by a fixed system because it didn't work properly. Northrop's first true flying wing design was for a passenger transporter. Avion became a victim of the Great Depression, but Jack Northrop succeeded in starting his own division of Douglas Aircraft. Northrop designed a number of successful aircraft for this company but lacked the time to make his dream come true. In 1938 he resigned.

A year later, helped by the fact that a role of the US was seen to be that of providing an arsenal for the fighting nations, Northrop Aircraft Inc. was established as a company for both designing and building aircraft. That company built and tested a range of different Flying Wings that were meant to be used as war planes, the first one being the N-1M with its characteristic drooped tips. It was a flying laboratory more than anything else. It proved Jack Northrop's theory and provided the knowledge and experience to proceed. The company developed the XP-56 fighter. It didn't perform as well as estimated and because it had propellers the Air Force lost interest since jet propulsion had announced itself. Northrop designed and built several promising bombers. The main problem for the Flying Wings was that development costs were too high at moments when these couldn't be met. Most Flying Wings were scrapped. There were also accidents. And there was competition. The B-52's success lead to the abortion of the Flying Wing project. However, during a flight by one of the planes, the YB-49, it was observed that the aircraft was difficult to trace by radar, but that fact was ignored. It did become a key feature for the B-2 stealth bomber, the development of which started in the 1970s, some 20 years after Jack Northrop withdrew from the industry just after having made a design for a large flying wing airliner. The B-2 was not a direct continuation of Northrop's work, although some of his research information came in handy. Northrop lived to see his dream come true in 1980. He passed away one year later.

A

B

C

D

E

F

G

Ever since the beginning of aircraft development Jack Northrop pursued his dream of building a flying wing. The Loughead brothers' F-1 (A) was the largest seaplane of its time with a wingspan of 74 feet. The S-1 (B) was the last plane built by Loughead. The company went broke. The plane performed well, but it was too expensive and there were too many 'Curtiss Jennies' in stock. In 1930 Northrop first attempted to build an almost tailless flying wing with his own company 'Avion' (C). Northrop's N-1M (D) was a flying laboratory. Different configurations of flight control surfaces could be tested. The fighter XP-56 (E) was coated with silver paint. It featured six gun ports, two of which were in the nose. It crashed (F) during landing after its fourth test flight. Two turbo jet engines provided thrust for the YRB-49-A (G). It first flew in 1950 and had a speed of 512 miles per hour at 40,000 feet. Unfortunately it was caught in a budget squeeze between two competing less advanced aircraft.

serve not only the environment but also the 'Investors' in education, research and development. It stimulates an economy based on creating high-grade knowledge and labour as an alternative for one based on a type of service that has already taken refuge in the low-wages countries.
When it comes to more efficient and environmentally friendlier air transport, we already possess the technology. For over 50 years there have been designers who want to get rid of the tailpiece of the traditional aircraft in order to eliminate a considerable amount of parasitic air resistance, dead structural weight and annoying loads in the fuselage. Distributing the payload in the loadbearing sections, as is customary for fuel, will lead to lower bending moments and torque and hence to lower running loads in the sheets and shells, which is the overture to a lighter structure. By integrating, or blending, the wing and the fuselage, the inner-wing fuselage acquires an enormous enclosed cross-section, which again substantially decreases the running loads from torque and bending in the shell structure, enabling an extra reduction in weight. This is a classical example of dematerializing by design, or replacing material by air for that matter.

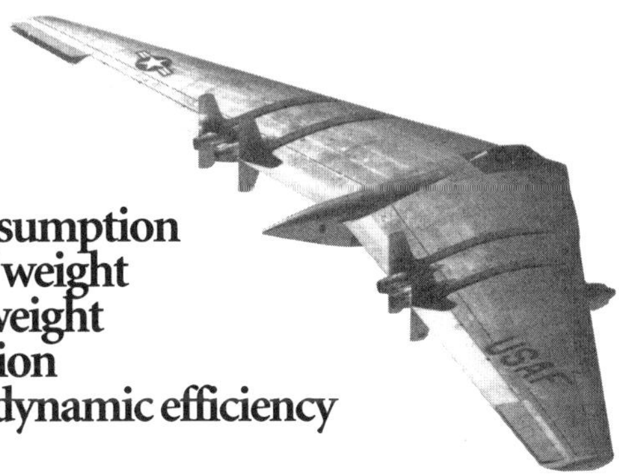

Blended wing:
27% less fuel consumption
15% less take-off weight
12% less empty weight
27% less propulsion
20% higher aerodynamic efficiency

More in 'Blending wings', page 108.

Left: A.M. Schwartz, Northrop's chief of structures, stands on the pilot's seat to get a good view from within the pilot's canopy. From this angle it can be clearly seen how the pilot's position is offset to the left of the aircraft's centreline. The co-pilot and bombardier sit on the right.

RUNNING DRY, RUNNING WET

Words from the song 'Kokoku' of American media artist Laurie Anderson: 'They say the dead will rise again. And here they come now. Strange animals out of the Ice Age.' Right, and what do we do with them? We burn their remains to get energy for transportation and there is an enormous amount of it. In the past more and more wells were discovered with a peak of over 40 million barrels in the mid 1960s. Since then the amount of discovered oil fields is steadily decreasing – about 10 million barrels per year now – in spite of improved research methods, and at the same time the rate at which oil is produced and consumed is increasing as it has always been. These days the level of production amounts to about twice the newly discovered stock.
Because of this, oil prices are on the increase and there are more conditions that contribute to price instability. The first is set by OPEC, which controls almost half the oil market. It has the power to close a few taps here and there, because of which prices will become higher. The second is that prices depend on stock estimations. If an OPEC country makes a high estimate of its reserves it will be allowed to produce more and get more money from banks. That holds true for private companies too. Recently Shell Oil Company got into financial trouble because it had overestimated its oil reserves. Something that is likely to become an oil price booster is the fast growing economy of China, with cheap labour and a sky-rocketing need for energy.
With strange dead animals reaching their limits it is only logical that alternatives need to be developed. The most obvious other energy provider for transport, including aviation, is hydrogen. This idea is not very new. The Russians flew the first hydrogen airplane in the 1970s. But recently interest in this way of powering aircraft has been renewed. Hydrogen can be used as a fuel directly or be turned into electricity. Both principles are currently under trial. Hydrogen has advantages. It burns clean. There is a side effect of nitrogen monoxide production (because of the burning heat), but this seems to be manageable and the contribution of water droplets and ice to the greenhouse effect may be less dam-

Towards the end of the 1940s interest in the tailless flying wing concept flagged, until this visionary idea by Jack Northrop and others became airworthy in 1988 in the form of the intercontinental Northrop B-2 bomber. The success and performance of this aircraft, this time artificially stabilized by virtue of active computer controlled operation, once more attracted the attention of researchers and developers, and a number of international (EU) and national (US) study programmes were initiated. Besides the flying performance, the placement of the engines and the pressurized cabin were also the subject of research. The results of these studies on paper were much talked-about.
In 1998 a team of NASA researchers concluded that both qualitative and quantitative performance improvements following from the transformation of a conventional 800-seater to the Blended Wing Body concept are unusual and unprecedented in the aircraft industry, and that all these improvements can merely be ascribed to the configuration, without even considering the latest findings in aerodynamics and structural design.
This shows that the specific resistance and the transport efficiency, as well as the parameters that quantify aircraft performance, can be substantially improved by newly designed aircraft. Moreover we can further improve the efficacy or the structural performance versus the costs, within the flying wing concept by applying new materials and a new design of materials and structures. These new opportunities will be further explored in the coming chapters.

aging to the Earth's atmosphere than the fumes produced by burning kerosene. The most important disadvantage probably is the volume it requires. Tanks need to be much larger. The development of light minimum volume storage systems is an interesting challenge. NASA is researching possibilities to build them from composite materials.

Real heating of the planet earth

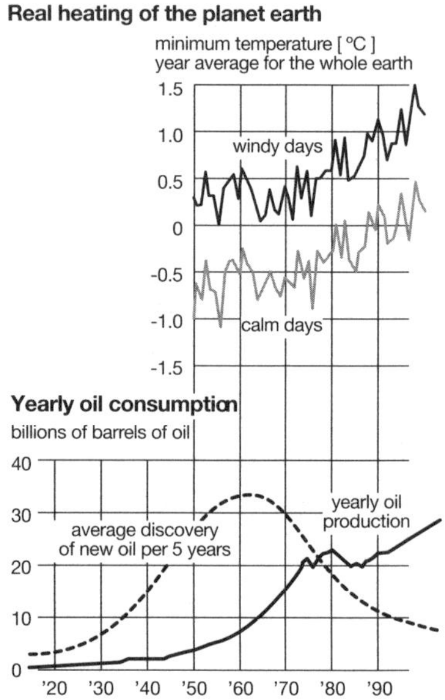

A real astronaut, André Kuipers, is turned into a cardboard model, whereas a Shell CEO called Philip Watts reckons a hero's outfit will help him convince an audience that there will be sufficient oil for years to come.

SPECIFIC DRAG. The graph provides an overview of the data in tables 1a and 1b. It shows the interdependence of drag (and any other kind of resistance) per unit weight and speed for different kinds of vehicle concepts. Speed of course relates to productivity, and low drag is always preferable. At the extreme left there is the barge, which is not very fast but doesn't suffer from drag at all. And it can carry a considerable amount of payload. The reputation of Zeppelins as an environmentally friendly form of transport goes down the drain since it needs a lot of energy just to elbow through air. Heavy cars like the Porsche Cayenne

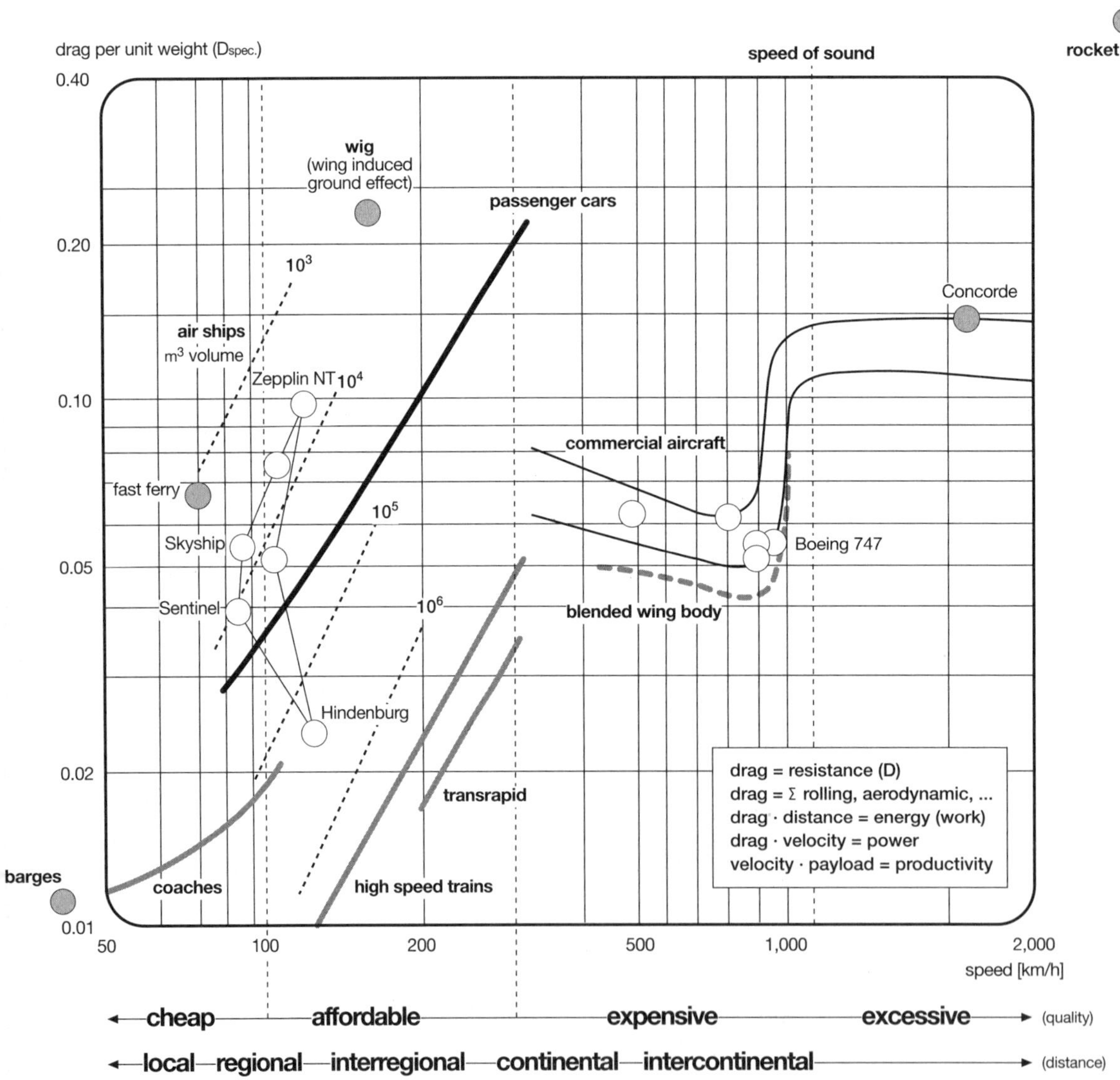

may provide a feeling of safety, but proves a not very adequate means of transportation from the efficiency point of view. On the right the blended wing body demonstrates a worthwhile increase in efficiency in comparison with current carriers.

A barge (bottom left) has to deal with mere viscous drag of water and the rocket at the top right doesn't have to conquer any drag at all as soon as it is in space, apart from what gravity does to it. Other means of transport fight various combinations of kinds of drag.

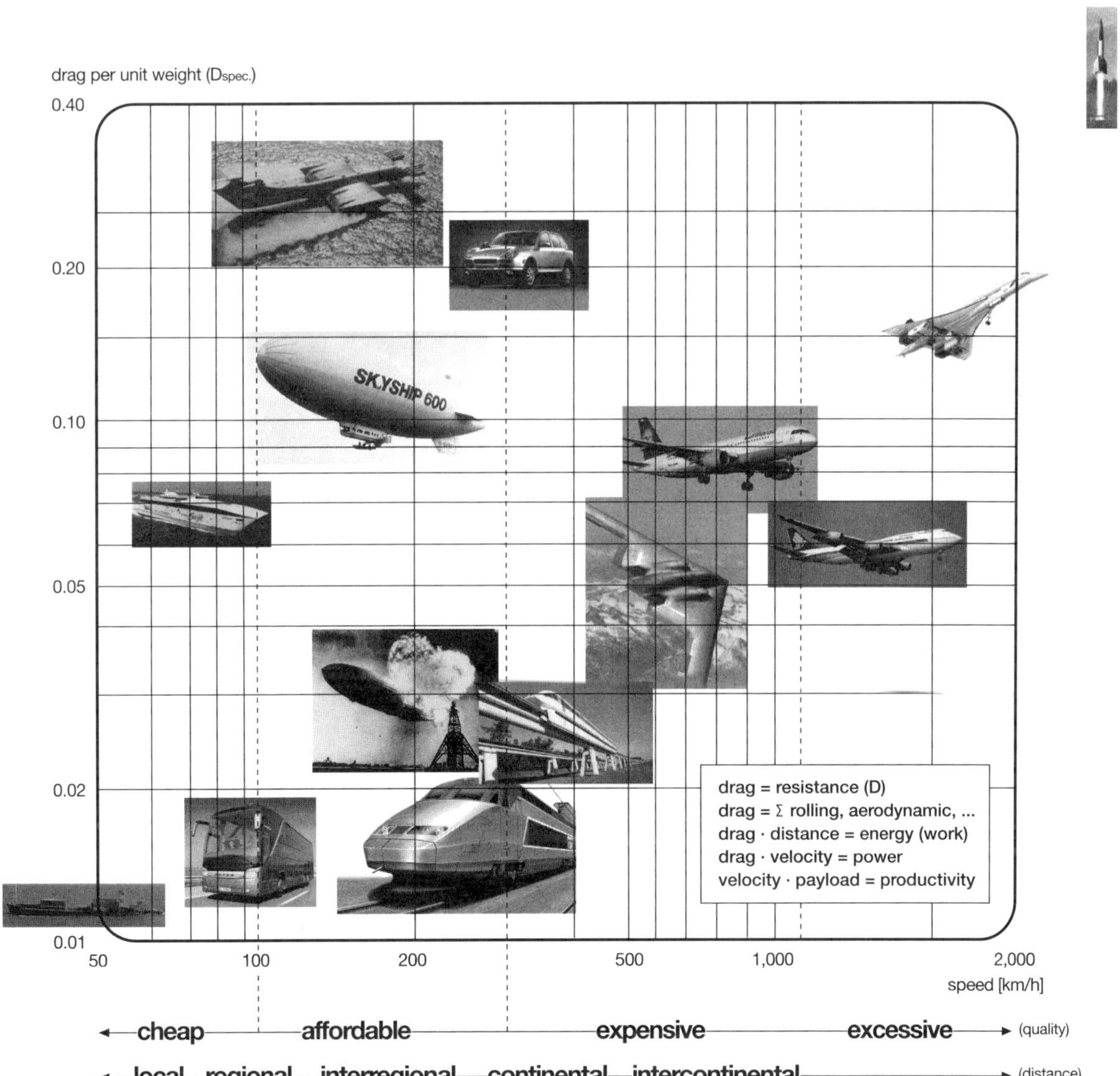

THE LAW OF BIG NUMBERS. Car sales people and their customers consider heaviness a sign of good quality. That is the main reason why heavy cars are more expensive per kilogram than light ones. And even high-ranking automotive CEOs tend to talk in a bantering tone about the effect of weight reduction on fuel consumption. This is how wrong they are: in the combination of road drag and friction, acceleration resistance and resistance due to driving uphill, which together add up to three quarters of the total resistance, the weight of the car is included three times. So fuel consumption depends on weight. As a matter of fact weight accounts for 0.7 litres of fuel per 100 kilometres times 100 kilograms on average. The simple implication is that a 100 kilogram weight reduction for just one stupid car leads to 0.7 litres reduction per 100 kilometres. A rough estimation demonstrates that even in the tiny Netherlands that would save 50 tank trucks per day, plus of course the diesel oil that they would require.

Car resistance equation

$$\Sigma D = \left[m_c \cdot g \cdot f_R + 1/2\, \rho\, v^2 \cdot C_D \cdot S + a\,(m_c + \Sigma\, m_{rot.}) + m_c \cdot g \cdot \sin\alpha \right] \frac{1}{\eta_{engine}}$$

$$= D_R + D + D_a + D_{grad}$$

D_R: ~35% — D: ~25%

operation (mean value): ~60 - 75% — ~40%

ΣD = total resistance
C_D = drag coefficient
S = cross section
D = air resistance
D_R = rolling resistance
D_a = acceleration inertia
D_{grad} = climbing resistance
η_{engine} = propulsion efficiency
m_c = mass of car
$m_{rot.}$ = rotating mass

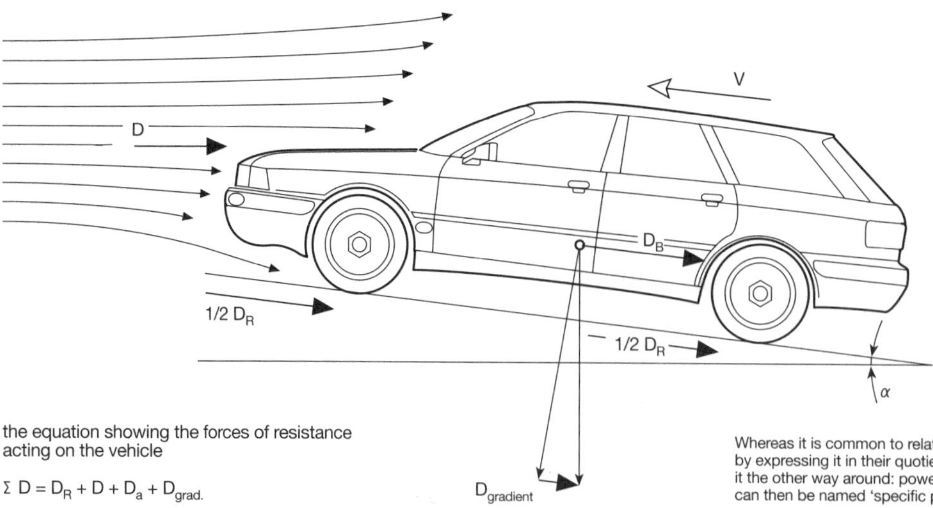

the equation showing the forces of resistance acting on the vehicle

$\Sigma D = D_R + D + D_a + D_{grad.}$

Whereas it is common to relate weight to power by expressing it in their quotient, we prefer to do it the other way around: power per unit weight can then be named 'specific power'.

WEIGHT BURNS FUEL. For different α's (α is fuel consumption per 100 kilometres and per 100 kilograms vehicle weight), the graph shows the linear increase of fuel consumption in relation to weight. If, for instance, your 1000 kilogram car drinks 7 litres of petrol in 100 kilometres, its α is 0.7. Which in this representation happens to be the α_2 line. It now shows you the reduction of fuel consumption when you somehow get rid of a couple of 100 kilograms of car.

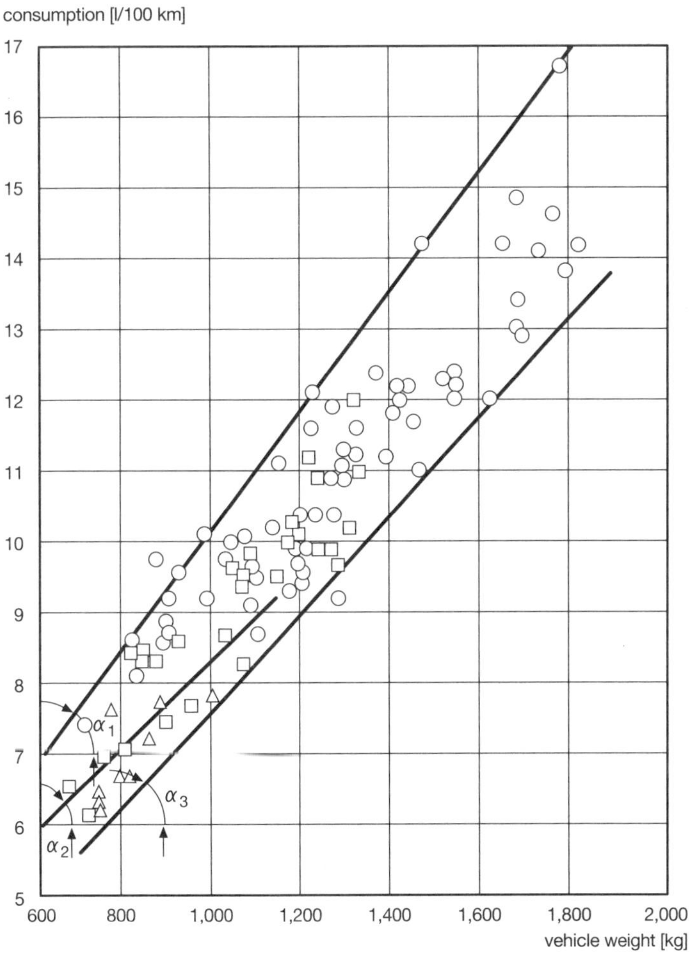

power to weight ratio [kW/kg] — α_2 : ○ < 14; □ = 14 - 18; α_1 : △ > 18

influence of weight

$\alpha_{1\ maximal}$ ~ 0.85 [l / 100 km · 100 kg]

$\alpha_{2\ optimal}$ ~ 0.35 [l / 100 km · 100 kg]

$\alpha_{3\ average}$ ~ 0.7 [l / 100 km · 100 kg]

TWO EXTREMES

Volkswagen seems to be somewhat indecisive about the issue of passenger car weight and fuel consumption, for this company produces both the very fuel-efficient Lupo and the Touareg, which is alleged to be one of the greatest petrol drinkers in the world.

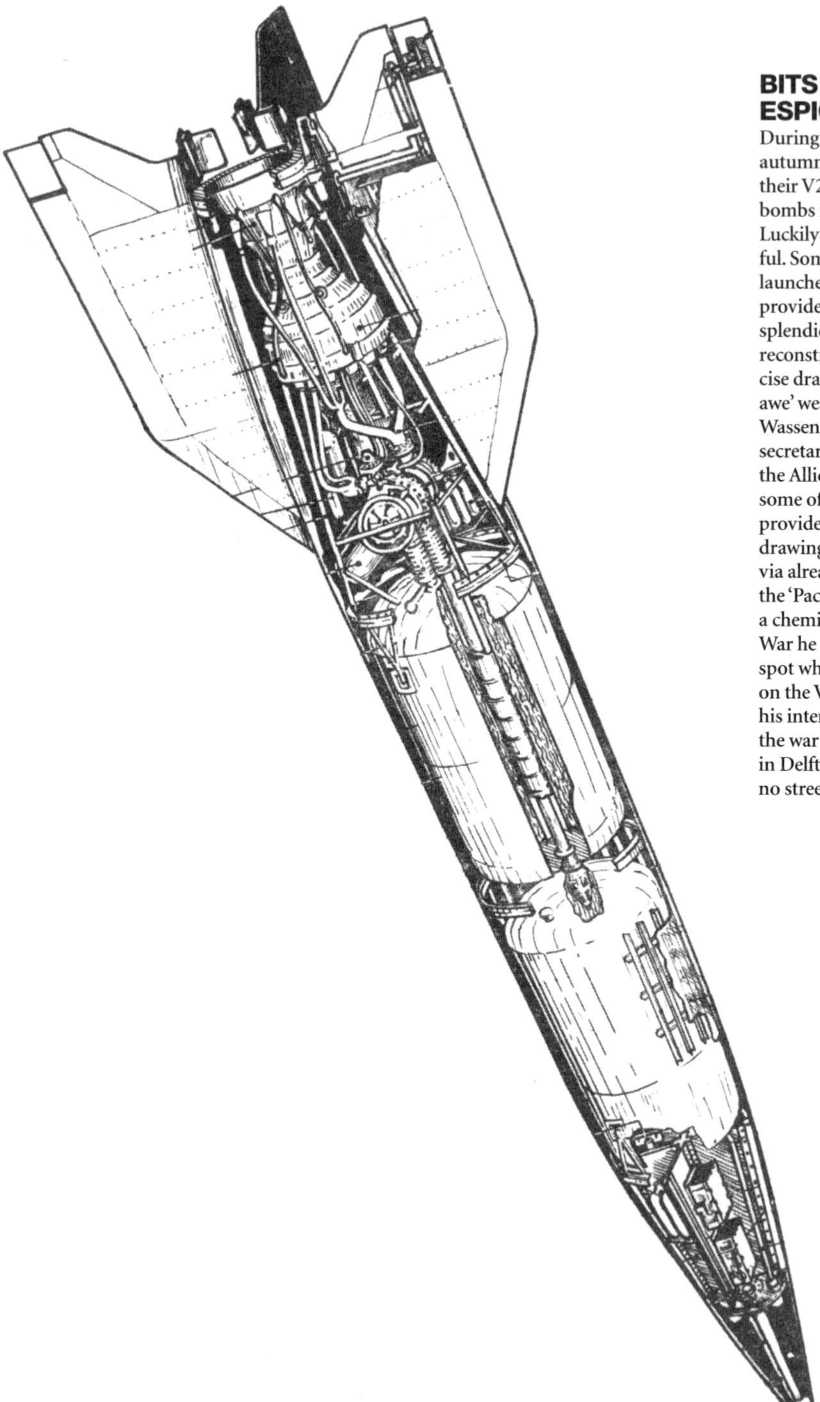

BITS OF ROCKET ESPIONAGE

During seven months, starting in the autumn of 1944, the Germans launched their V2 ('Vergeltungswaffe Zwei') rocket bombs in The Hague to destroy London. Luckily not all the missions were successful. Some exploded as they were being launched. Scattered V2 bits and pieces provided J.F.W. Uytenbogaart with a splendid but dangerous opportunity to reconstruct the concept and make precise drawings of this early 'shock and awe' weapon in the attic of his home in Wassenaar and secretly – not even his secretary knew – send the information to the Allied Forces. A group of messengers, some of whom were caught and killed, provided him with the broken parts. The drawings were transferred to 'London' via already liberated Antwerp through the 'Packard' network. Uytenbogaart was a chemist and before the Second World War he had worked in Berlin, near the spot where Werner von Braun worked on the V2's predecessors. That had raised his interest in rocket technology. After the war he became a chemistry professor in Delft. He died in 1964. Up until now no street has been named after him.

The standard work that Uytenbogaart wrote during the war, together with J.M.J. Kooy (Ballistics of the Future – V-weapons) about orbit mathematics has become a precious collector's item.

GROWING THRUST. The Wright brothers are an icon of aviation because of their first motorized flight, but they founded their commercial success on the 'motorized' part of their fame. They developed a type of piston engine that drove the propellers of many airplanes all over the world. With their ingenuity they were able to develop light piston engines cast from aluminium and magnesium and optimize air-cooling for each cylinder by arranging them in a circle around the crankshaft into what became known as the radial engine. The Wright Cyclone, recognizable through its typical air outlets was the choice for most aircraft until gradually jet engines took over. The Wright company still exists and is now called Curtiss-Wright. No longer a specialized producer, its activity dissipated into metal technology service providing, a fate that awaited many metal product manufacturers.

Now a propeller works best where pressure is already high and it does very little with temperature, whereas a jet engine tries to bring up both pressure and temperature at higher altitudes, where pressure is insufficient for propellers to turn into thrust. Over the years jet engines have become better and better in this, because of improvement in materials, technology and efficiency. Whereas classic engines reduced the volume (increasing pressure) by a ratio of 4 to 1, now the most modern engine can manage 12 to 1. Ironically, because of the reduction of the number of turbine blades and the increase in diameter, jet engine fans are beginning to resemble propellers again. Moreover one of the two engine types that will propel the Boeing 7E7, the General Electric GEnx, has composite rotor blades. Rolls Royce attempted it before in the 1970s, when they developed a new engine for the Lockheed TriStar, but they failed and almost went broke because of metal mimicking rather than authentic composite design.

One other archetypal engine type needs to be mentioned, a very old one: the rocket engine. It starts out from low temperature and no pressure at all, and generates thrust just by burning fuel. It first became successful in the Second World War in the shape of the dreaded V2 flying bomb. Typically a rocket engine has to carry both reactive substances to generate thrust. This is an advantage in outer space, but not in atmospheric conditions. There have been attempts to develop intermediate thrust generators. The 'scramjet' can use air at very high altitudes, but it is inefficient where propellers and jet engines thrive. Ideally there would have to be an engine that can change its characteristics according to altitude. Not easy.

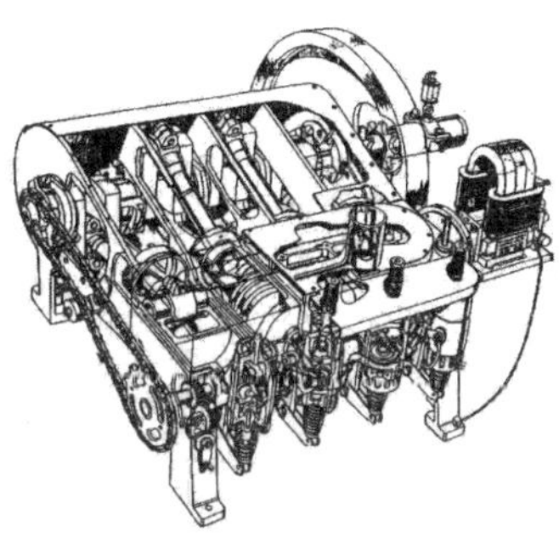

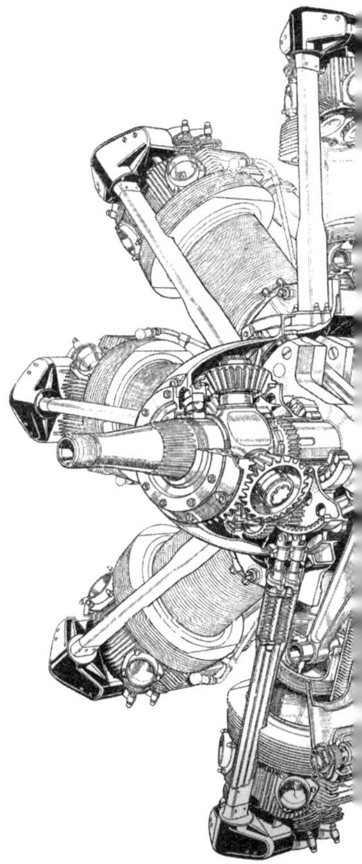

The Wright brothers are famous for their motorized flight in 1903, but their success in business stems from the aero engines they developed, particularly the Wright Cyclone radial propeller engine, which provided thrust to many pre-jet aircraft.

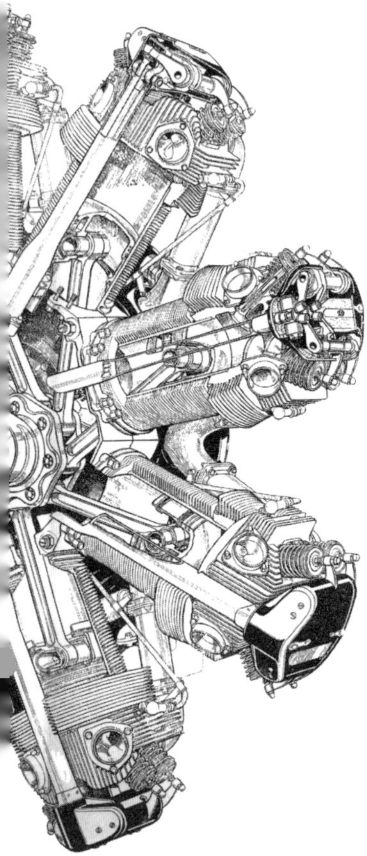

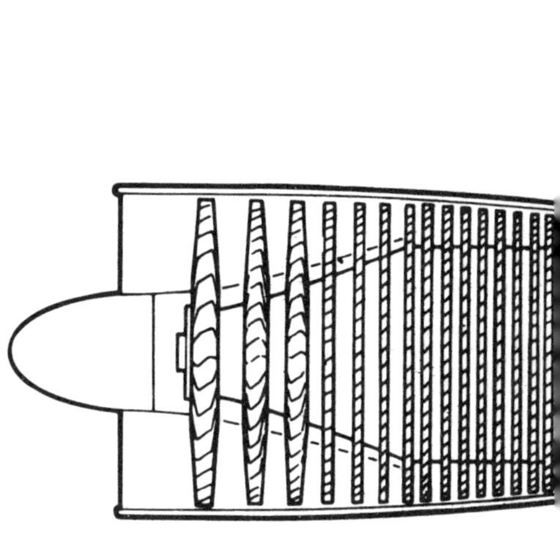

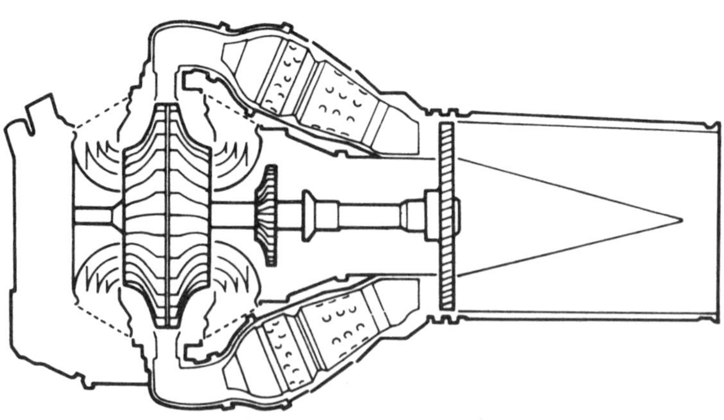

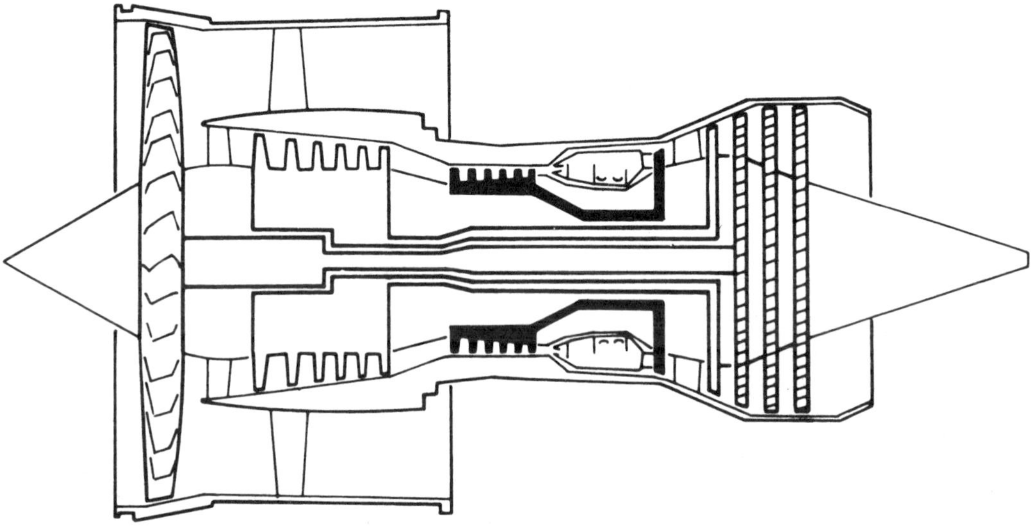

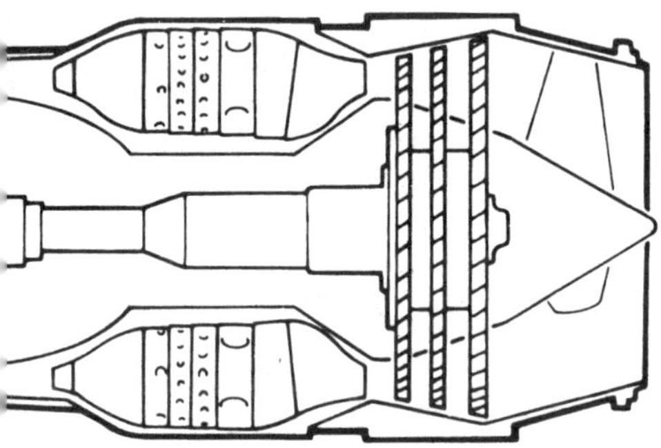

The jet engine has become more efficient compressing more air with its turbine. The fan, ironically, is starting to almost look like a propeller again.
Left 1940s, middle 1950s and right an example from the 1970s.

Past inertia

Air travel did not evolve from a clever down-to-earth technical invention to the current complex organization of human flying behaviour along a simple line of increasing sophistication. The process encountered contradicting disturbances caused by fierce and fragile industries, safety considerations, economical ambitions, political interests, war, technological challenges, and tourist courses to conquer the fear of flying. Apart from structure and payload, there is another important determinant for the weight of an airplane. This factor is commonly known as 'emotion'. It isn't something you can put on a scale as such, but its influence on the current state of flying is considerable. The foresight of the exhilaration you would feel if you could look down on your own home from high up in the sky caused airplane pioneers to find a way to really take off. However, emotions that accompany power, honour and prestige, as well as unfounded feelings of being right, or the fear of doing something wrong or breaking rules, have hampered the development of aviation to the point that currently it has come to a near halt.

Between 1900 and 1902 the Wright brothers built a number of gliders. Some of those crashed. The brothers wisely used failures to reconsider their calculations. Left: Otto Lilienthal just before take-off in 1891, before the Wrights, but no engine.

A little over a century ago mankind saw its first powered flight, which was performed by Wilbur and Orville Wright on 17 December 1903. These brothers' unique and trend-setting enterprise, their skills to develop, build and, later on, commercialize load-carrying flying machines, is a rare example of entrepreneurial engineering. They understood

STEEP THRILLS

Burt Rutan has broken the 18-kilometre mark. His SR-71 cruises at over 24 kilometres. He wants space travel to go commercial: 'There is WAY too much red tape in the government to do anything. As much as I hate to say it, I think the Chinese have a much better chance than we do. If they lose a couple of astronauts, no big deal. No one is going to get voted out of office because of it. Everyone who goes up in the shuttle knows that there is a chance that something bad will happen and they might not return. With 120 missions flown, two ending in the loss of craft and crew, I would take the less than two per cent chance of dying just for the opportunity. How many men have died in the exploration of the ocean? Well I am of the opinion that above us is one large ocean... and we are just sticking our toe in the water... every once in a while the water nips us.'

what they were after, and their work involved wind-tunnel experimenting with lifting devices and full scale tests of major structural components. More importantly they developed a lightweight 36 horsepower engine (with 4 cylinders in line and a fuel injected carburettor) that drove two propellers using bicycle chains. Moreover the Wrights developed the principles to control flight movements. As a matter of fact their twistable wing tips for rolling and heading were the last remains of the ancient dream to truly mimic birds.

The Wright Flyer's engine represents an important step away from even the clumsiest bird.

The Wright approach to the design and development of an airplane as a system, which consists of independent sub-components and subsystems, earlier advocated by Sir George Cayley (1773-1857), meant nothing less than the beginning of an era. It was the start of a worldwide development of materials, a wide range of different aircraft concepts, configurations, engines, control tricks and equipment inventions, all supported by rules and tools. During the first 30 years after the first motorized take-off, most of the building blocks for reliable aircraft were developed.
The first attempt to materialize flying machines was based on the use of natural materials such as yarns of silk or flax woven into dense stiff fabrics or 'taffeta', supported and prestressed by 'fingers' of young sprouts of birch or ash. It provided the lightest possible wing structures, similar to those of birds and bats and, of course, the archetypal sketches of

A BIT OF ART HISTORY

Flying was first and foremost interpreted as a dream come true. That fact was strikingly illustrated by the way in which it was received at the onset. Dutch journalist and historian Jan Blokker wrote a wonderful article about it in the Dutch newspaper 'De Volkskrant' on the occasion of the 100th birthday of motorized flight. It begins with 'Pioneers of aviation have broken the fallow grounds between heaven and earth, between death and life, for us. In a hundred years something great has been accomplished. Aviation is no longer above, but within our power.' Here follows a very brief excerpt: 'The beginning of aviation was strongly related to art. In 1909 Franz Kafka attended an early air show in Brescia, Italy and was immediately fascinated. He took a course in mechanics and wrote an essay on the Planes of Brescia. The cubist painters Braque and Picasso adored the Wrights, and the bicycle-making brothers must have been flabbergasted when, during their visit to Paris, they were lionized by the art elite. Picasso even called Wilbur 'mon cher Vielbure'. Everywhere composers, painters and choreographers expressed their feelings of congeniality with the men that really turned their dreams into reality. Stranger, however, was the total absence of commerce. No entrepreneur recognized the potential of the Wrights' primitive machinery. Maybe this is not that strange if you consider the slow evolution during the first years. With trains things moved a lot faster. No sooner had the first steam locomotive pulled five carts with a total of 70 passengers at a speed of 8 miles an hour over a distance of 15 kilometres, than vast numbers of prospective exploiters were ready to start a new business. In Paris Heinrich Heine wrote: "The most elementary notions of time and space are fading away. Already I can smell the German lime trees from here and the North Sea waves are lapping around my front door." He would never have written something like that for the Flyer. It was in the air for only 59 seconds, covering a meagre 260 metres. Even Julius Caesar was faster with an entire army when he conquered Gaul.'
(Ed note: A fully equipped (33 kg) Roman soldier could walk 30 km/day)

The death of France's most beloved flying ace Guynemer as painted by Lucien-Hector Jonas in 1918. The work was entitled "Heroes of the air don't die".

Leonardo da Vinci. Weavers, upholsterers and shrine workers became the first builders of bird mimicking machines. This most artisanal technique was applied to build the glider of Otto Lilienthal, the first successful attempt to mimic soaring birds but without the engine. In the next stage, bicycle and coach manufacturers built frames and truss constructions out of steel wire, with spruce for compression rods and beams loaded in bending using primitive industrial processes. Firmly pre-tensioned linen was used to upholster spruce ribs, spars and girders to form aerodynamic air foils. This is what the Wright brothers did.

The Mosquito could fly at high altitudes and was the fastest and stealthiest airplane in World War II, although it was made out of old-fashioned wood. Most Mosquitos ended up as fuel for fireplaces.

Next, high performance plywood laminates allowed the construction of rigid stressed skin wings and fuselages, as in the first Fokker and Lockheed Vega planes. These wooden shell structures were glued together with natural starch. Some time later the change was made to synthetic glues of phenolic resins in order to overcome temperature and moisture sensibility. Although the Fokker, the Lockheed Vega and the fast and stealthy DeHavilland Mosquito, the first and quite successful composite airplane, outperformed most of their contemporary metal counterparts, plywood was soon to be replaced by aluminium. Sheet parts were attached with rivets, according to the same method used in iron shipbuilding, to plate structures similarly riveted together. The introduction of aluminium was purely ideological. Applying any other 'old fashioned' material was akin to heresy. On the other hand, because of this one-sided attention to alu-

MUTUAL REFUTATION

Wood was successfully replaced by aluminium in the 1930s on the basis of four naive assumptions disguised as arguments. 'Metal doesn't burn and wood does', ignoring the key role of fuel, 'metal parts are cheaper to produce on an industrial scale', overestimating the scale, 'metal is more durable', not yet knowing about the key role of fatigue, and 'metal is stronger and thinner and therefore lighter', not realizing that thin metal parts tend to buckle under compression. Apart from this belief, the predictability of mechanical properties linked metals to 'science' and 'progress' and made wood 'old-fashioned'. The arguments were purely rhetorical and not based on research, let alone experience. The idea that metal airplanes were 'incombustible' proved to be a lie. Metal airplanes were perfectly capable of burning, melting through the heat of burning gasoline. Out of eight metal German Junkers JU46 bought by the American mail services, four were destroyed by fire because of this. Weight is a very complicated issue since it depends, by its very nature, on structural efficiency, which in its turn varies with the nature of the loads that components have to bear. It took enormous effort to build lighter planes in metal. The first metal bomber was simply too heavy for its destructive payload. Buckling, based on structural instability, appeared to be a dominant design criterion. Stressed skin concepts could handle shear by diagonal tension. However, skin under diagonal compression stress were not weight competitive at all because of this buckling phenomenon. Attempts to control it in aircraft resulted in complex structures locally stiffened with lots of extra parts and with abundant riveting to attach them.

Problems of buckling and durability of aluminium structures like intercrystalline corrosion and fatigue, were scien-

tifically investigated and either solved or made controllable. Consequently, metal airplanes proved to be much more expensive to produce than wooden ones. The reason for continuing originated from the firm belief that mass production of intermediate parts and well organized assembly would reduce the production costs in due time, as it had done in the automotive industry. Even in the 1950s there was this idea that airplanes would one day become a consumer product, a family vehicle that would take you anywhere. This idea was also nourished by the fact that many young American men had got a flying license, after a brief training to become a pilot in WW 2.

The wrong initial arguments in favour of aluminium were rapidly overtaken by other, more substantial but rather unexpected advantages such as resistance to the stress caused by the pressurization of the fuselage at high altitudes, and the ability to shape it according to the rules of aerodynamic smoothness at high air speeds. In this respect the capability of the metal air frame changed large scale air transport into a success. Now, after 70 years of improvements and optimization of aluminium alloys, parts manufacturing and assembly, which by today's industrial standards can be considered artisanal, the traditional metal 'semi-monocoque' structure has reached the stage of consolidation and even decline of the technological life cycle.

Now the same four arguments that were unjustly used to advocate the change from wood to metal aircraft can be used rightly to promote the change from metal to composites for future aircraft structures. As an extra, composites feature superb crash energy absorbing capacity as shown in Formula I racing. Moreover composite structures can be made to last and they are sustainable as well.

46

minium, tremendous and unforeseen progress was made in stressed skins, supported and reinforced by stiffeners and frames.

Three early full metal aircraft in chronological order: Junckers 52, Douglas DC-2 and the Boeing 307 Stratoliner.

It implied the overture to the true revolution of industrially produced all-aluminium stressed skin aircraft such as the Boeing 247 (1932) and the highly successful Douglas DC 2 and 3 (1933). The ability to build airtight pressurized cabin structures, such as the Boeing 307 'Stratoliner' (1937), was the start of 70 years of evolution into today's very efficient high velocity transport system, using almost the same aircraft configurations, scaled from small to mega and ultimately equipped with very reliable high bypass jet engines. Yet, the very same religious attitude that gave aluminium the edge over any other option, did hamper alternatives like composites. Today, the aluminium aircraft concept together with its manufacturing infrastructure is approaching its technological limits. Every opportunity to improve efficiency and weight/drag ratio, and compete in a stressful market, is seized upon, even if this implies strange and inefficient solutions or politically dividing up aircraft structures between different stakeholders in dubious ways. Design agility suffers from conflicting interests.

The shape and concept of this personal portable pressure cabin holds a promise for blended wing pressure cells yet to be developed.

LOVING AND HATING WINGS

Dreams of other people can be quite confusing, especially if they have almost the same name. In the 1930s there were two brothers in Germany, Reimar and Walter Horten. They were involved in model airplane building and, just like Jack Northrop, they had this dream of a tailless airplane. It was to be just this superb flying wing. From the middle of that decade onwards they started to experiment with real manned flying wings. They were made according to what at that time was conventional technology: a wooden or tubular steel frame, covered with veneer wood. There was one exception: the skin of the Horten Va was made of Trolitax, an epoxy resin. This plane crashed into smithereens. Reinforcement hadn't been invented yet, for resins anyway. There were some accidents. One of them killed the test pilot, but how this could happen remains a mystery. He could have made radio contact, or bailed out, but he didn't. Their first aircraft was a glider. Later models were motorized with propellors, turbojets and even jet engines. The planes flew quite well and Reichsmarshall and former pilot Hermann Goering became interested. On one occasion in autumn 1944 at a demonstration he demanded that model H VII flew with one engine turned off and there were no problems. Nevertheless the Horten brothers' flying wings never rose above the status of training aircraft. At the end of the war a centre section of the last model Ho 229 was found by the Americans and stored in Silver Hill, Maryland. Horten designs are still flying, but they are scale models. Not long after the Horten brothers, in the 1950s there was an American with exactly the opposite dream, except that the two dreams turned out to be very much alike. Bill Horton wanted to build wingless airplanes that could be driven like cars. Automobiles in those days looked a bit like airplanes anyway, so his idea was not that strange. His designs basically consisted of a wing section with vertical tip plates at each end to provide directional stability and extra lift. They were primitive lifting bodies. Some prototypes were built and could actually fly. What remains of Horton's dream is a wonderful illustration of a typical fifties ideal: a plane for everyone.

Fortunately, sciences in aeronautics and lightweight structure engineering are at the brink of the next revolution: the change from all-metal technologies to hybrid structure technologies.

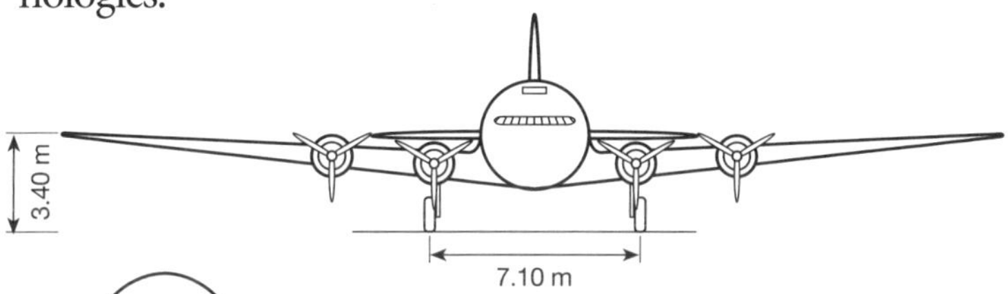

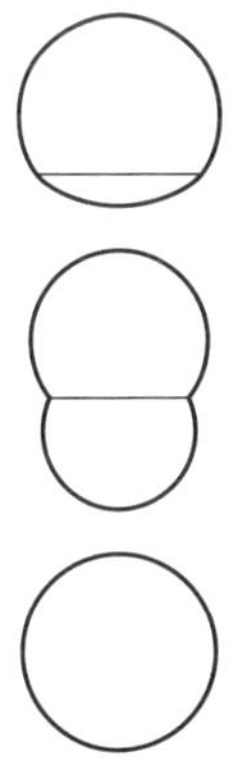

The cross-section of the fuselage determines stress distribution. The Boeing 307 'Stratoliner' was the first passenger aircraft with a pressure cabin.

42

These last-named are based on textile-reinforced polymers which are locally 'blended' with metals. In the near future a tremendous morphological leap will take place: mechanically assembled solid plates and slender beams will make way for flexible integral three-dimensional fibre structures using ancient textile manufacturing techniques. They only have to be 'frozen' into ultramodern multifunctional shapes with the help of plastics.

Ironically, current metal designers use almost the same arguments against composites as the ones that were used to support the replacement of wood by metals: 'we have a poor understanding of anisotropic mechanical behaviour', 'its behaviour is hard to predict and its properties vary too much', 'composites burn', 'they suffer from fast degradation', and more. But for modern synthetic composites these arguments are no longer valid. These days the mechanical and physical behaviour of solid structures of composite plates

DODGY DESIGN

Enforcing compulsory, non-constructional constraints can render the mechanical life of a structure very difficult or even impossible. Even today, aircraft design is still not the integral rational process one would expect from an advanced industry. Rather it is a ritual, dominated by marketing and geopolitical arguments which sometimes govern early choices, unavoidably leading to insoluble problems in a later stage. The requirement to integrally mill with expensive and prestigious machines the wing skins of a small 50-seater developed in a Far Eastern country which shall remain nameless, resulted in unprecedented wall thicknesses, ranging from 11 mm at the root to 1.5 mm at the tip, the usual range, in for example the F27, being 3 to 4 mm at the root to 0.7 mm at the tip. Thickness is heavy, so to get the weight of the wing back into competition, it was decided to reduce the sectional plane of the torsion box, the sleeve that provides resistance against torsion. This decision resulted of course in reduced torsion and also bending stiffness. Together with a poor finish due to the limits of milling, this led to an unacceptable life expectancy of only 9000 flights.

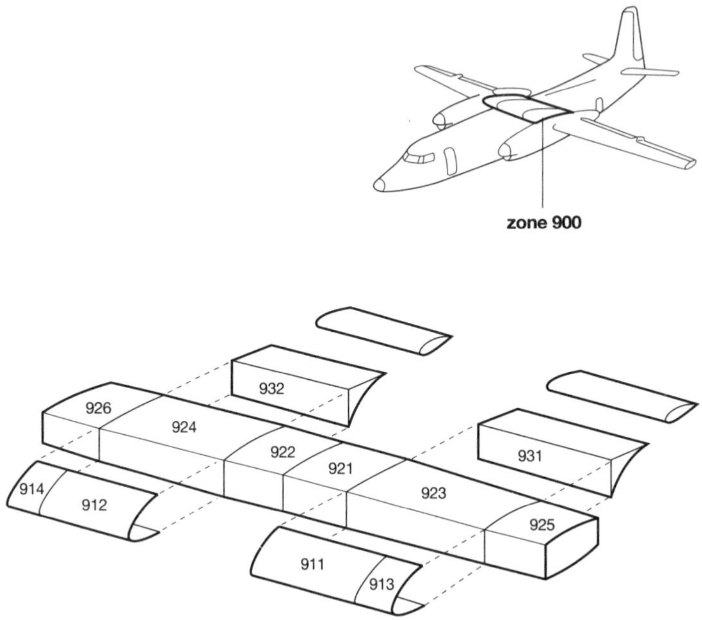

Fuselage cross section

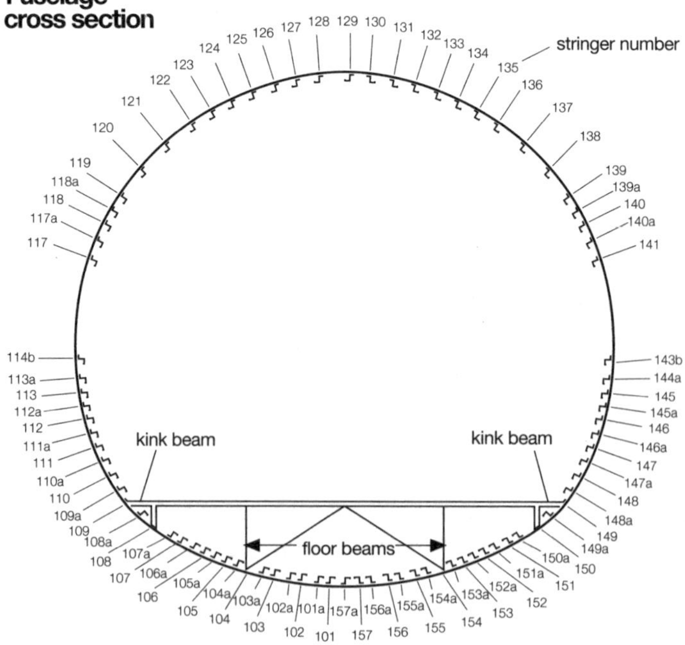

The cross-section of a tube-shaped pressure cabin can be optimized for optimum stress distribution and minimum weight because of that.

and shells are perfectly predictable and well understood. Moreover modern composites can be made to be fire resistant and extremely durable with the proper recipe.

THE RIGHT TO FLY

A number of questions will have a major influence on the future transport of people and goods and on the extent to which potential improvements of means of transport can be reached. That is, if policy makers are willing to recognize their importance.

The most important of these are:

- Will the number of people with access to high speed transport increase with the growth of global prosperity, or will this type of transport become the privilege of a restricted market and a limited range of goods? It's a matter of price. Will the market consist of people who are willing to pay a decent sum of money, or will it be a mass market of football supporters, sun tanners, cucumbers and frozen chickens?
- Will air transport be able to meet the changing demands of a growing and increasingly affluent world population? The demand for local and, even more so, regional and continental transport is on the increase.
- Can we limit the damaging effects of air transport to the environment and public health? Now we're back at being willing and able to pay that decent price.

The difference between theoretical and true cross-sections is considerable.

Two KLM stewardesses and a co-pilot dance on the roof of a Boeing 747-400 at Guatemala City Airport, for a birthday or something. It was the stunt of the year 1997. They were severely reprimanded.

KINK BEAM. In the cross-section of a typical aluminium pressure cabin structure the 'kink beams' support the cabin floor. They also define the floor as a border surface between the upper and the lower parts of the fuselage. Depending on the cross-section design, the floor has to deal with different kinds of load when the fuselage is actually pressurized. For structural efficiency tension load is of course preferable, mimicking the tension distribution in two adjoining soap bubbles. The 'floor' in between them can remain very thin and the principle is applicable to composite pressurized structures.

Loads in floor structure

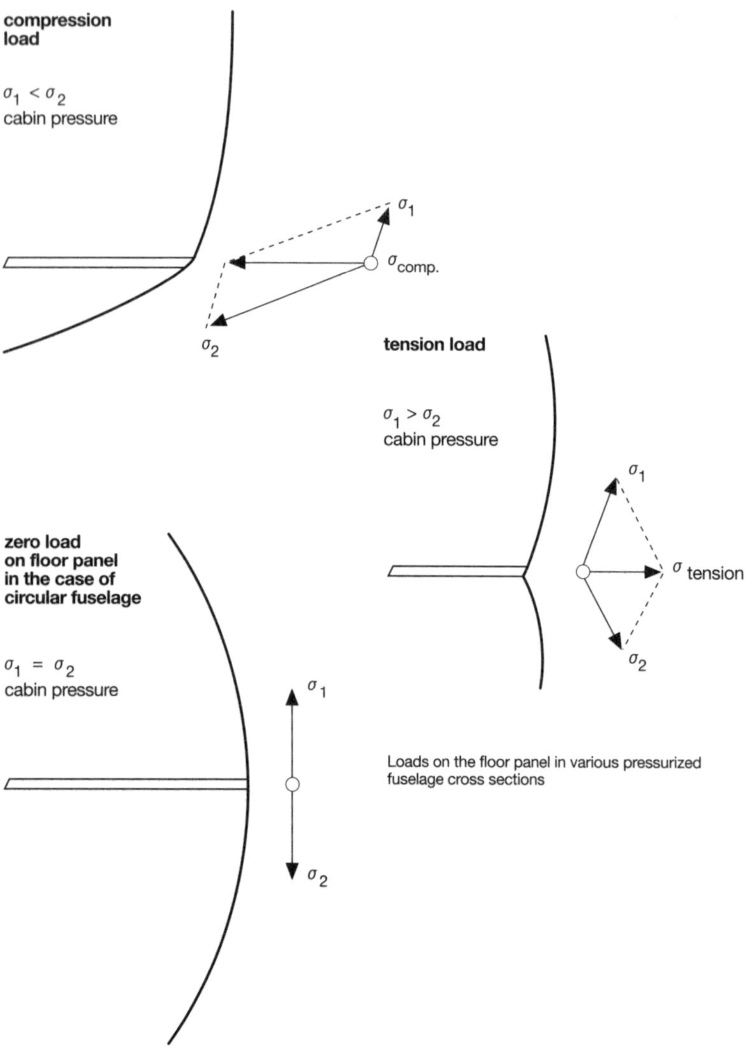

Loads on the floor panel in various pressurized fuselage cross sections

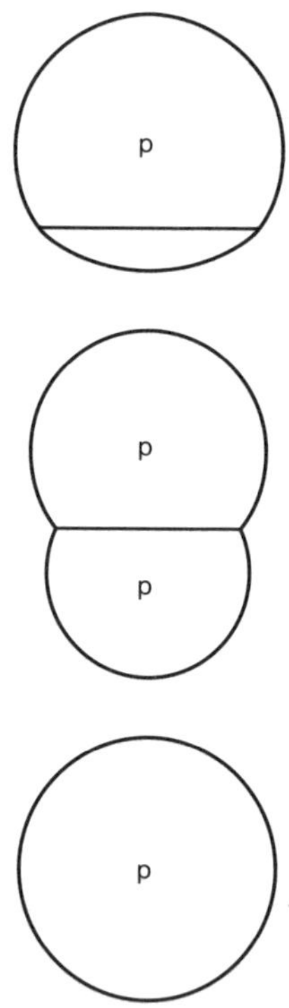

Tension stress, as always, provides optimum conditions for weight reduction and soap bubbles provide the perfect example.

Radii of two coalescing bubbles (r and s) and their common surface (t) obey a simple relationship

$$\frac{1}{s} = \frac{1}{r} + \frac{1}{t}$$

$p_2 > p_1$

$\sigma_1 < \sigma_2$

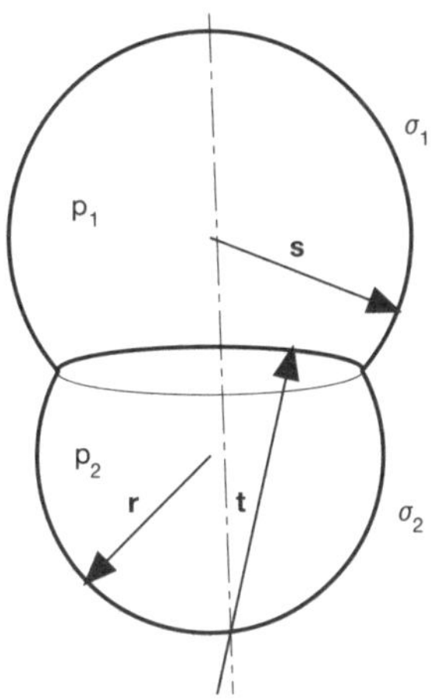

Meridian of an isotensoid shape

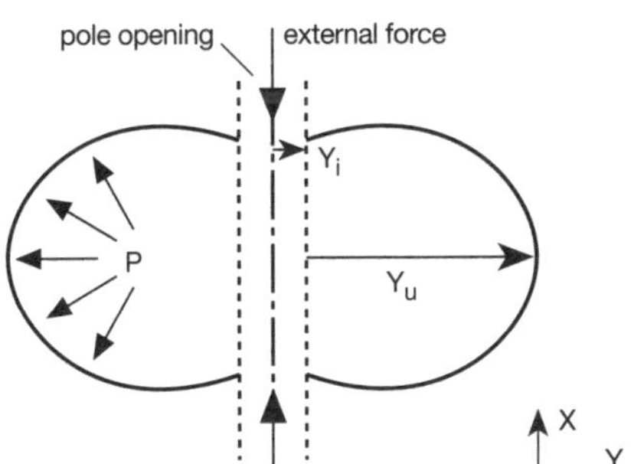

$$\tan^2\alpha = \frac{\sigma_r}{\sigma_a} = 2 - \frac{R_n}{R_\mu}$$

$$n\,F\cos\alpha\,\cos\beta = P\,\pi\rho^2 + K$$

Sankey diagram 1930

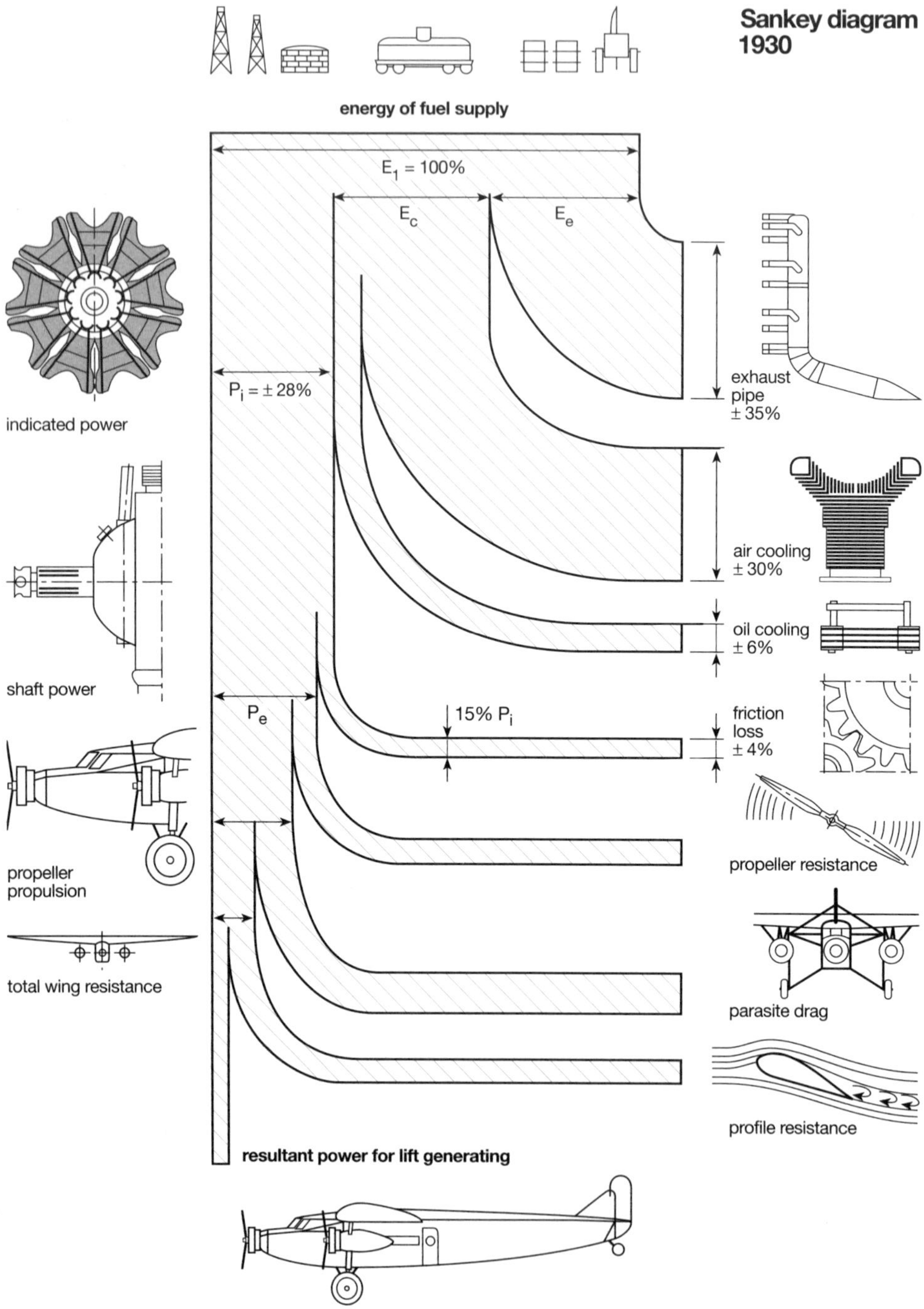

POWER DISTRIBUTION. The efficiency of an airplane engine is typically around 30 per cent. That is the part of the energy that is actually delivered to the propeller shaft. The rest is wasted in cooling, the exhaust, and conquering drag and friction. This 30 per cent is about the same for car engines and jet engines. However, jet engines deliver best under different conditions. To maximize efficiency you play around with temperature and pressure.

Sankey diagram jet engine 1950

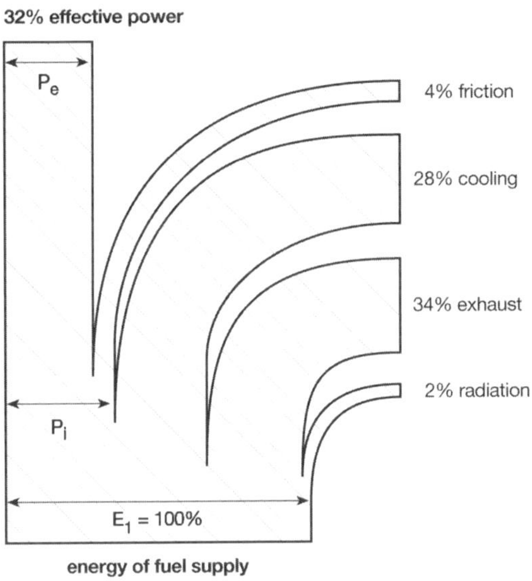

General Electric's new GEnX turbine promises an improvement of useful power to 40 per cent.

	Lift	Thrust	Control
THEN	 Bird wing	 Man	 Bird (torsion)
	 One wing	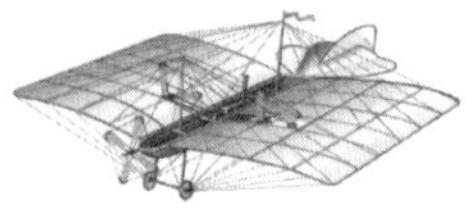 Steam propeller	 Torsion
	 More wings	 Otto propeller	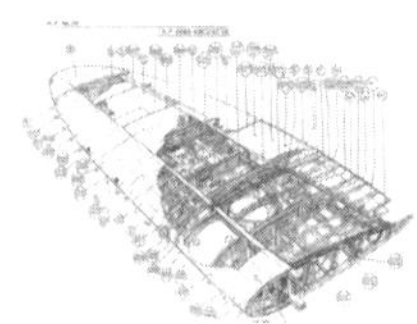 Mechanical
	 One wing	 Jet	 Fly by wire
	 Optimization	 Optimization	 Wireless
NOW	 Birdwing	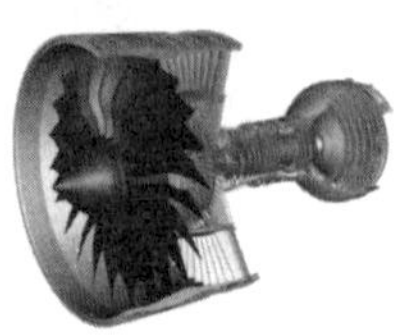 Extreme bypass	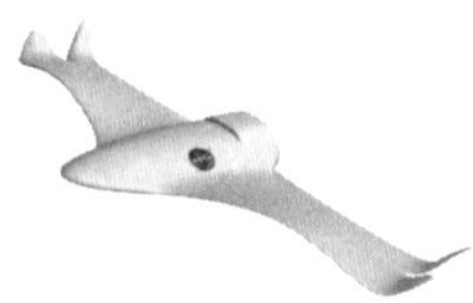 Decentralized Border layer suction

Payload

One man

Couples

Bombs & Mail

Passengers

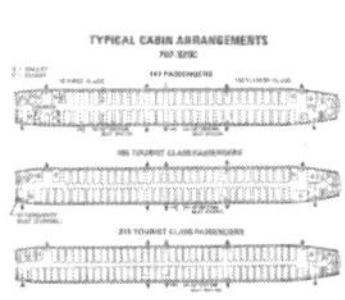

Mass transport

Service

None

Modest

Lite

Spartan

Economical

Luxury

Sir George Cayley split up the airplane in functional systems into 1853. This subdivision has become a standard thereby hampering air transport development.

Gravity rules

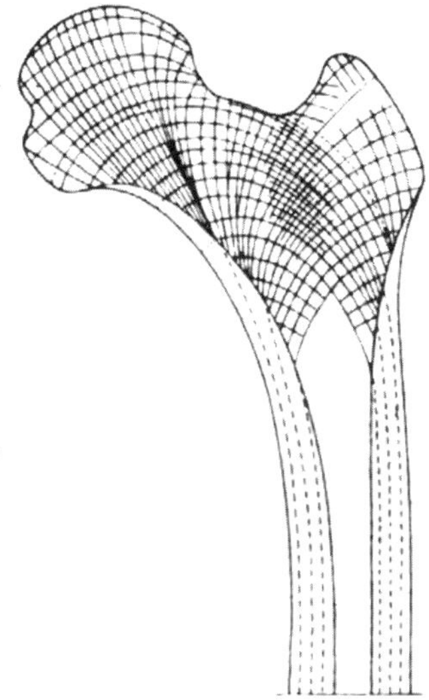

The Michell structure resembles the stress trajectories of the cancellous bone found in the human femur, and many other bones for that matter.

Flying is conquering gravity, or at least part of it. A century of aviation and related fields of technological development has left its marks on our understanding of airplane building principles. Regardless of material or shape, the art of constructing lightweight structures thrives on an intelligent distribution of mass, and thereby the points of gravity application, over space. Even the food industry can instruct us here. A cross section of just one potato crisp demonstrates maximum performance using a minimum amount of material. Millions of consumers are being satisfied with air surrounded by a filler material that carries the required taste and texture and, most importantly, produces a nice sound when it cracks. Here we have a high ranking technology resulting in a high quality product that consists of almost nothing. Crisps are an excellent metaphor for lightweight structures. Material properties per unit surface area should be optimized, for example in order to achieve the highest possible strength and stiffness for the overall structure. Therefore it is essential to avoid discontinuities in stiffness distributions and load paths, the trajectories along which loads are transmitted. Harmony is the key here. Dissonance gives rise to bewilderment. Light structures have to be light through and through, from skin to bone.

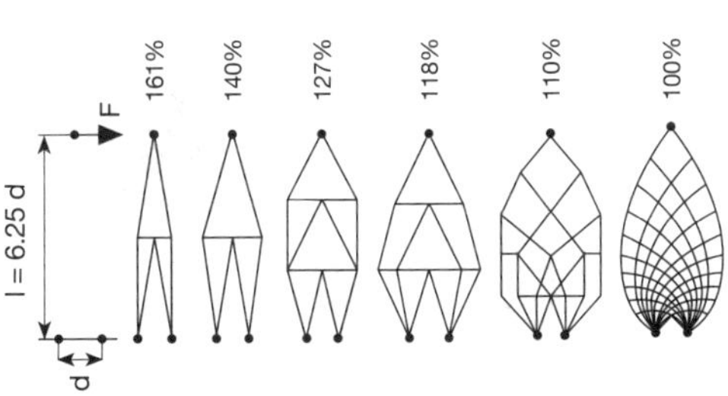

Replacing bars by more and thinner ones eventually leads to the much lighter Michell structure, which nevertheless has the same strenght.

There are a number of basic rules for supporting a maximum load with a minimum of material in a theoretical static world, without deformations and vibrations or damp-

ening foam. To start with, a structure needs as many 'natural' load paths as possible, with a minimum of diversions or dead ends, or in techno-posh words: the topology of load paths should be optimal. Next, if possible, try to replace beams that are under pressure with systems in which tension forces are predominant. Think, for instance, of diagonals in frames and of 'tensegrity' structures, as devised by Buckminster Fuller. Furthermore 'parasitic' bending and torque, forces that are the side effect of inefficiency in structural details such as joints and profile cross-sections, need to be avoided. When bending loads cannot be prevented, maximize the distance between the main load paths, or the 'critical moments of inertia'. Similarly, in the case of torque in thin walled closed cross-sections, try to maximize the enclosed surface, the 'polar moments of inertia'.
The following examples exist in that same academic idealized, static environment. Nevertheless they demonstrate the considerable power of structural design when weight reduction carries a high priority: In all frame structures we can optimize pressure and tension loaded beams in such a way as to achieve a weight reduction of as much as 60 per cent, said Michell and Maxwell in the early 20th century.

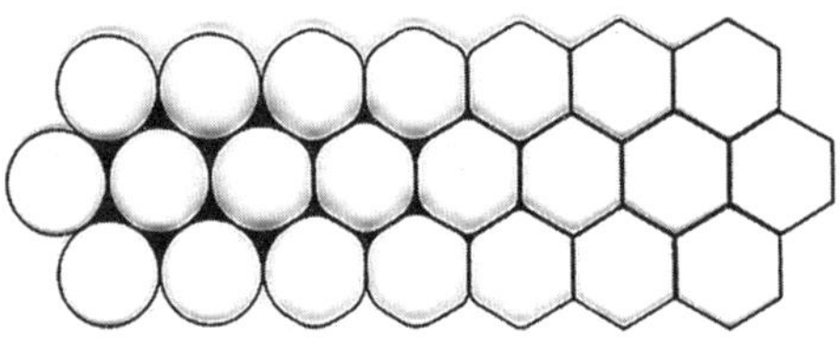

Stacking spheres and making them share their skins leads to the minimum energy structure of foam.

This optimazation is very common in nature and therefore highly desirable in light spatial structures. For optimally reinforced barrels and columns it is a natural derivative of the design and applied loads. Now take a diagonal in a 7075-aluminium frame structure. It carries its load performing as a tension loaded beam, thickness t, with 75 per cent less material than the aluminium tube, with the same wall thick-

FOAM SANDWICH

Some structural principles carry the name of food which in this case got its name from an English lord. Sandwich shells will become increasingly important. In a sandwich structure two relatively thin skins are made mechanically functional by inserting a thick and airy core. From a strictly mechanical point of view a sandwich structure is heavier than the classic stiffened and 'empty' shell structure, but as a system the sandwich concept has the potential to become lighter and cheaper. There is no further need for stiffeners. Because of that no mass is lost, which results in a higher stiffness value per unit skin surface. The result is that the number of stiffening and strengthening elements needed for solid sheets will be greatly reduced. Stiffeners cannot be avoided in places that suffer from high loads, such as wing-fuselage connections, landing gear attachments, doors and hatches. Sandwich simplification of the structure will yield enormous effects on production and maintenance costs, not to mention the elimination of corrosion issues. Sandwich systems feature better acoustic insulation as well, especially for the lower frequencies, due to low skin mass. The same effect occurs at higher frequencies, thanks to the higher bending stiffness per unit area of the sandwich. Add to this the effects of absorption and internal damping when cellular elastomer-like core materials are being used. And let us not forget true leakproof thermal insulation, which can be achieved by applying the right core materials. Heat loss of aluminium (200 W/m2K per mm sheet thickness) can be reduced by as much as 98 per cent simply by applying a plastic foam core of only 1 mm thickness. As a rule, foamed nonflammable polymers, with sufficient chemical resistance, strength and stiffness, will suffice. Sandwich structures have the natural ability to satisfy a great number of requirements by their integration of various functions. This explains why there are so many examples of sandwich structures in nature, for example the human skull, well, any skull really. Integration of functions in different materials implies their redistribution. So a new balance between segregation and integration has to be established, over and over again.

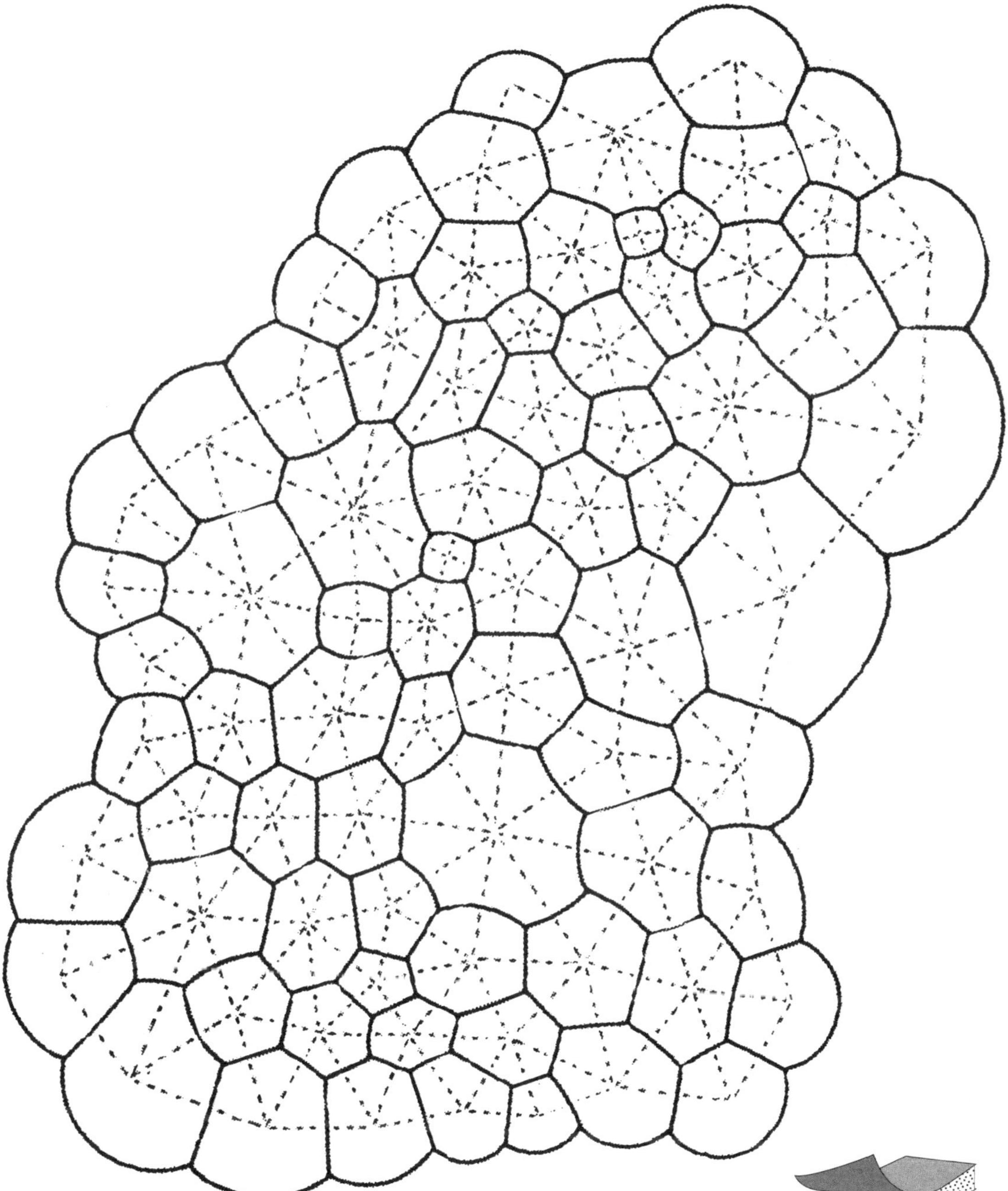

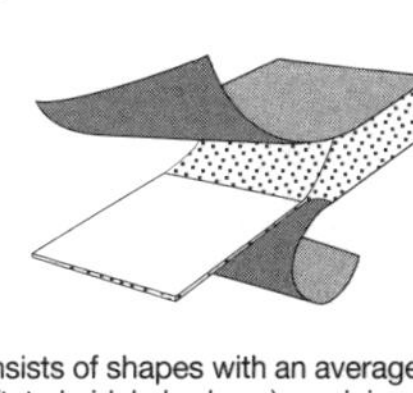

Foam consists of shapes with an average of 14 faces (tetrakaidekahedrons), a minimum energy structure in the sense that stress is optimally divided in space.

U.S.
MAC

If the fuselage of this freighter had been designed in fibre-reinforced plastic the picture would have had an entirely different atmosphere, because of the changed balance between segregation and integration. More functions would not have been visible as separate additions.

ness t, loaded under pressure, provided the length L of the tube is at least 70 times larger than the characteristic diameter d. If the resistance against either bending or overall instability under pressure dominates, we can even reduce the weight by 80 per cent by improving the design of the cross-section. In that case you could say metaphorically that the profile cross-section has evolved from machine building, via mechanical engineering to aerospace engineering.

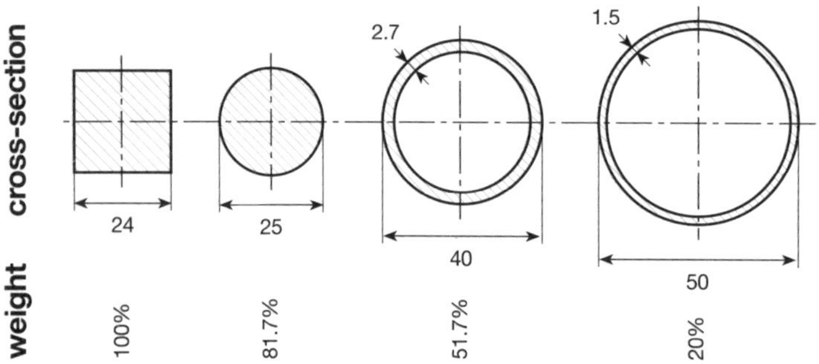

This range of cross-sections have the same bending strength characteristics but the amount of material, and therefore the weight, is successively less.

As said earlier, for sufficiently strong and stiff aircraft structures, the potential of possible improvements in metal structures has virtually been exhausted. As in bridges, semi-metaphorically speaking, only the transition from metal to composite structures and the change of the morphology of materials and constructions can get us any further. Replacing massive sheet and profiles with fibre-metal laminates, and maybe even fibre and textile reinforced polymers, seems to give the opportunity to create hybrid structures with better mechanical properties. However, straightforward replacement more often than not is also the least intelligent option. This is where composite turns into that darkest alternative material of all: 'black metal'. If we use 'black metal' in this way, not only do we introduce danger, such as infinite lengths of free edges and curves that are subject to peeling loads, but we also ignore the completely different behaviour of these new materials in the presence of load concentra-

COSMIC MUFFIN

The first passenger carrier to fly at high altitudes was Boeing's 307 'Stratoliner'. The difficulty of developing a cabin that could maintain the pressure of about 3600 feet at 14000 feet was quite an achievement, remembering that the cabin consisted entirely of a riveted structure. It is not easy to find ways to prevent air from leaking out when you do it for the first time. It took two years, 1937 to 1939 to get the airplane certified. Ten were built in total. Five of them were used in the war as freighters. They had Wright R1820 radial motors.
The plane also caught the eye of mystery millionaire Howard Hughes. He had one customized for his own use, the Stratoliner SB 307B NC 19904. The cabin of this particular plane has been transformed into a houseboat named 'Cosmic Muffin'. It is for rent somewhere in Florida. One complete and flightworthy Stratoliner remains, the Clipper Flying Cloud. It is on display at the National Air and Space Museum at Dulles International Airport Washington DC.

81

tions. No more slowly growing visible cracks. With composites we enter the world of explosive brittle fracture or creeping delamination.

Replacing materials rather than redesigning structures also ignores the fact that replacing a cheap tried-and-tested component with something new, such as a composite, can only be successful if we can offer more value for money. Take for example the part of the aircraft that is most expedient in terms of material replacement, the pressure cabin. Traditionally this is a prismatic one-cell structure and in flying wings it is a multi-cell structure. This airtight pressure cabin is not only the part of an aircraft that, being the heaviest, has the most difficulty in conquering gravity, but is also the most expensive and the most complex because of the sheer number of components. All these aspects can be improved upon by means of an integral redesign, in which balanced integration and segregation of all functions are key factors. The starting point for the choice of new materials, structural concepts and production techniques should be an intended improvement of the price-performance ratio of the system as a whole.

Currently a large variety of sophisticated computer tools for structural design and analysis is being developed. Most of them are based on the use of linear and nonlinear finite elements, which in a sense implies considering a structure as if it consisted entirely of mathematical foam. With advanced displacement, vibration and optimization codes almost all technical design problems can be tackled.

FLYING PIPES

Backgrounds are sometimes difficult to hide. Whereas the Wright brothers used bicycle transmission parts, the German airplane manufacturer Hugo Junkers (1859), who was also a pioneer in building full metal airplanes, started out as a developer of gas engines and later geysers and gas heaters. His early patents for metal wing structures betray his affinity with pipefitting.

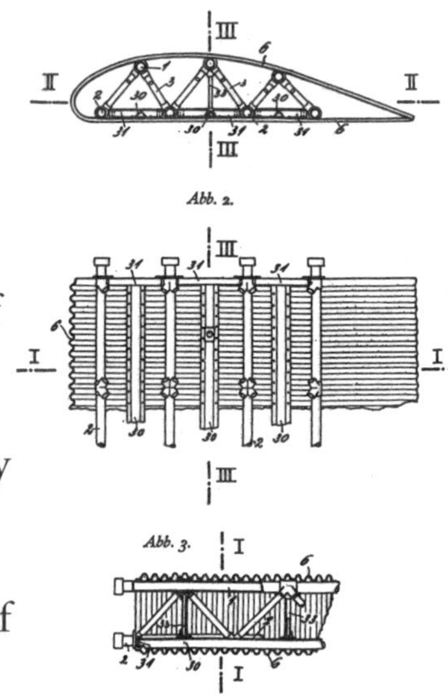

Metal structures are quite complex, since virtually every function needs its own special element, all introducing its own attachment difficulties. Metallically speaking, system integration is a matter of adding rather than sharing.

20 oz
Enjoy
Coca-Cola
CLASSIC
Coca-Cola

However, to prevent designers from having blind faith in 'toolbox results', a quick insight into mechanical capabilities is indispensable. Therefore 'classical mechanics' are still important, even for the design of composite structures, which by their very nature behave differently in different directions. A great number of classical analytical equations 'describing' the elementary behaviour of structural elements under different loading conditions are quite useful. In 'mass and gravity dominated' mechanical problems, big scale cable structures for example, they help select materials and optimize structural efficiency. This family of pure tension loaded structures is big: all the way from one-dimensional cable structures to tents (their two-dimensional fabric equivalents), and from rapid rotating parts to pressurized vessels. By using very simple analytical equations, some ordinary engineering and a smattering of material knowledge, many kinds of tensoid structures, from civil superstructures to small scale pressure vessels, can be quantified and qualified for efficiency. The big potential of carbon fibres, as pultruded rods or dry strands, is obvious. When stiffness (expressed in Young's modulus E) per unit weight is a crucial criterion, for instance in pre-tensioned cables and stays, it is even greater. Of course if the design of anchoring carbon fibres is not tackled properly, steel designers still have the edge. For researchers and developers, however, applying modern composites can be very challenging and offer openings to new paradigms for structures and products. In addition it is essential to get the same insight into and feeling for structures and loads in the world of beams, plates and shells. This is a simple matter of time and hard work.

IMPERSONATION DESIGN

One of the classical mistakes often made with a new material, is to leave the original structural and functional design largely unchanged. Often if aluminium is replaced by carbon reinforced resin without considering structural change, what you get is jokingly called 'black metal design'. Such mistakes are not limited to one kind of replacements. Over and over again new developments are strongly influenced by mimicking the devious certainties provided by predecessors. That can be the only reason why a Coca Cola can is shaped like the successful bottle.

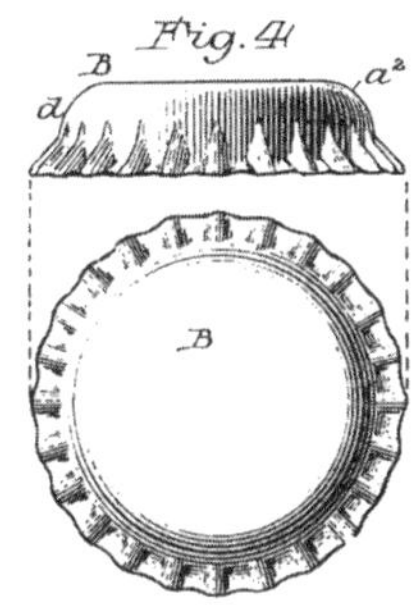

59

In 1891 William Painter, an American of British descent who lived in Baltimore, Maryland, invented the crown cap. It had a corrugated-flange edge, which was possible because of very simple and clever metal technology, difficult if not impossible to replicate in any other kind of material.

The difference between suspension and flying seems a modest gap to bridge.

諸國名橋奇覧
飛越の堺
つりはし
前北斎為一筆

GRAVITY FOR MODELLING. Catenaries formed by gravity are natural minimum length curves or surfaces with respect to span and sag. Huygens and Leibniz (1696) were the first to describe them mathematically and from ancient times to recent history they became the archetype for architects (in ancient Greek archi-tectón means weaving supervisor) to make roofs for their spaces. The Spanish architect Gaudí used catenaries to model three-dimensional buildings under the assumption that compression is the opposite of tension. He made a mistake. Whereas a slender structure can withstand tension stresses virtually independently of its cross-sectional material distribution, a supporting construction may fail at low compression stress levels due to instability caused by insufficient bending stiffness and strength. As soon as bending and shear stresses occur in Gaudí's creations, mathematical integrity is undermined by the mechanical laws of deformation, calling for measures 'to keep up the fabric'.

Gaudi, possibly the most functionalist architect ever, cleverly used gravity induced catenaries to develop his complicated arch structures.

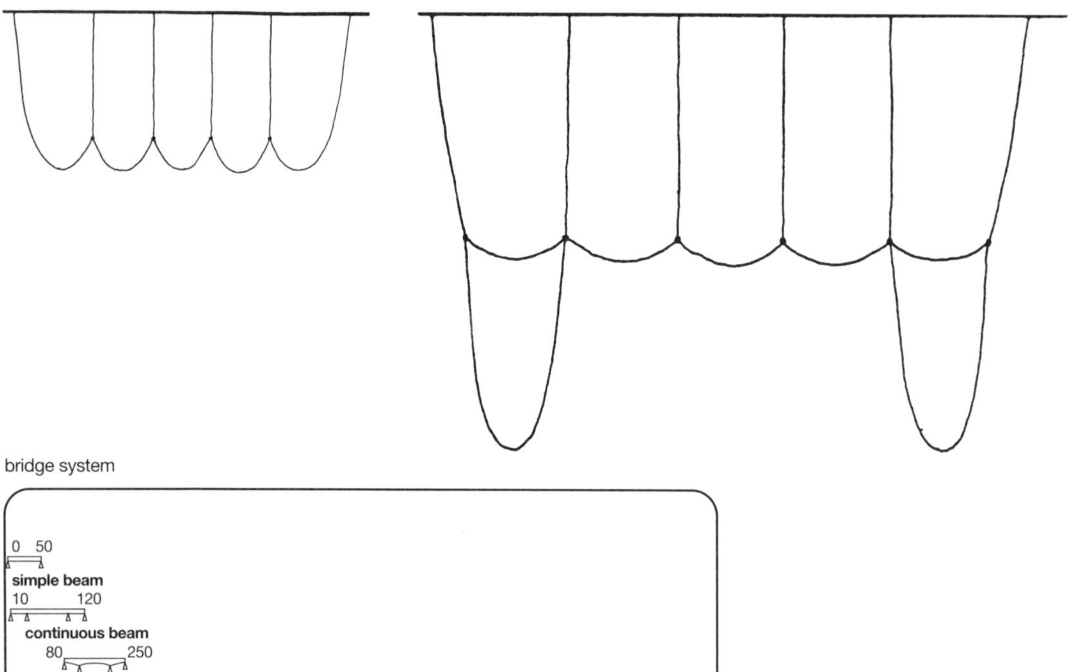

bridge system

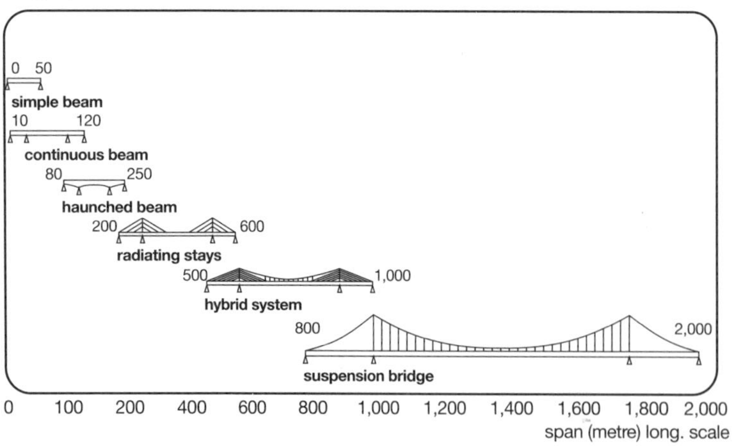

The more the structure of a bridge depends on pure tension, the longer the possible span. If it is more than two or three kilometres, the material that takes up the tension forces starts to become important. Steel is no longer good enough.

GRAVITY'S SHAPE. Spiders don't know it themselves, but the million years success of many of their families is based on their capability to spin and weave fibre webs with unbeatable properties. Stays and catenaries probably are the oldest of all applied mathematical and mechanical structures. Most webs are 'woven' radials and tangentials. When loaded by wind, raindrops or heavy impact the web structure of stays and catenaries shows extreme deformation capabilities without breaking.
All over the world skilled tribesman as well as civil engineers are still inspired by the web's 'building blocks' to build all kinds of bridges, and bridges, especially big ones, are like aircraft, since their weight determines their performance.
Suspension bridges are particularly beautiful examples and they feature superb material efficiency. In these big scale cable structures performance is not determined by using lots of material, but rather by specific material properties per unit weight. When neglecting the loads of moving air and vibration, these properties determine the final performance, which is the maximum span between supports.
Bridges are of course built to allow heavy 'traffic' to pass. Relate the load to be carried W' to the weight of the deck W" and the weight of the supporting cable structure W''', and you get the specific design load (y), which is the 'payload' per unit weight of the structure. With it you can calculate dimensions without knowledge of absolute volumes and weights. For catenary bridges the traditional magnitude for the design load (y) is about 0.5, which is a result of the decision to equalize the weights of the deck, the cable structure and the maximum traffic weight. For tension structures dominated by mass, using slender prismatic rods or cables, the classical theories of elasticity offer elegant expressions for strength and stiffness. Even for flexible cables without a single natural undeformed state in idealized

CRANDALL'S CENTENNIAL JUMPING ROPE.

Playing jump-rope is a way to turn a gravity catenary upside down with the help of centrifugal force.

A. Pre-tension cable properties

	σ [10^6 N/m^2]	ρ [10^3 kg/m^3]	E [10^9 N/m^2]
steel	1860	7.8	200
carbon fibre	2400	1.7	140

carbon fibre properties specified for UD-pultrusions with 60% fibre volume

maximum span [m]; $\beta = 5.36$

specific design load y	**steel** stayed	**steel** suspension	**carbon fibre** stayed	**carbon fibre** suspension
0.25	160	1190	1330	7120
0.33	230	1470	1840	8830
0.50	**350**	**1980**	**2870**	**11860**
....	....	....	....	....
inf.	1830	5990	14890	35580

geometrical conditions, simple strength and stiffness equations can be derived for predimensioning purposes.
When stiffness factors and loads other than weight are neglected, the bridging performance of stays and catenaries can be expressed by the following equations.
When typical modern pre-tension cable properties (failure strength ($\bar{\sigma}$), specific mass (ρ) and Young's modulus (E)) are introduced (table 1), the maximum span (S) versus specific design loads (y) can be calculated. We have to bear in mind that not the intrinsic material properties determine maximum span, but rather the efficiency and reliability of anchoring (expressed by the safety factor b = 5.36). The calculated curves and data show the mechanical superiority and efficiency of suspension bridges over cable stayed ones and the potential of tremendous improvements when the bundles of high strength steel rods are replaced by bundles of HT-carbon epoxy pultrusions (vf> 60%) (graph 1). The cost increase of the material must be brought into balance with either a larger span or a reduced number of supporting towers. And they don't need painting over and over and over again.
Carbon cables could yield a similar improvement over steel ones in conventional suspension bridges. Under ideal circumstances and comparable cable anchoring performance, the maximum span could increase from a maximum of 2 km with steel cables to at least 10 km with carbon ones. This would be more than sufficient to bridge the Straits of Messina (3.3 km) and Gibraltar (8.8 km).

$$S_{cat} \leq \left\{ \frac{\lambda}{\beta\left(1+\frac{1}{y}\right)} \right\} \cdot \left\{ \frac{\bar{\sigma}}{\rho g} \right\}$$

λ = 1.325 for optimal catenaries

$$S_{stay} \leq \left\{ \frac{1}{\beta\left(1+\frac{1}{y}\right)} \right\}^{2/3} \cdot \left\{ \frac{96\ \bar{\sigma}^3}{\rho^2 g^2 E} \right\}^{1/2}$$

Max. span of classical suspension (S_{cat}) and cable stayed (S_{stay}) bridges

Suspension bridge versus cable stayed bridge

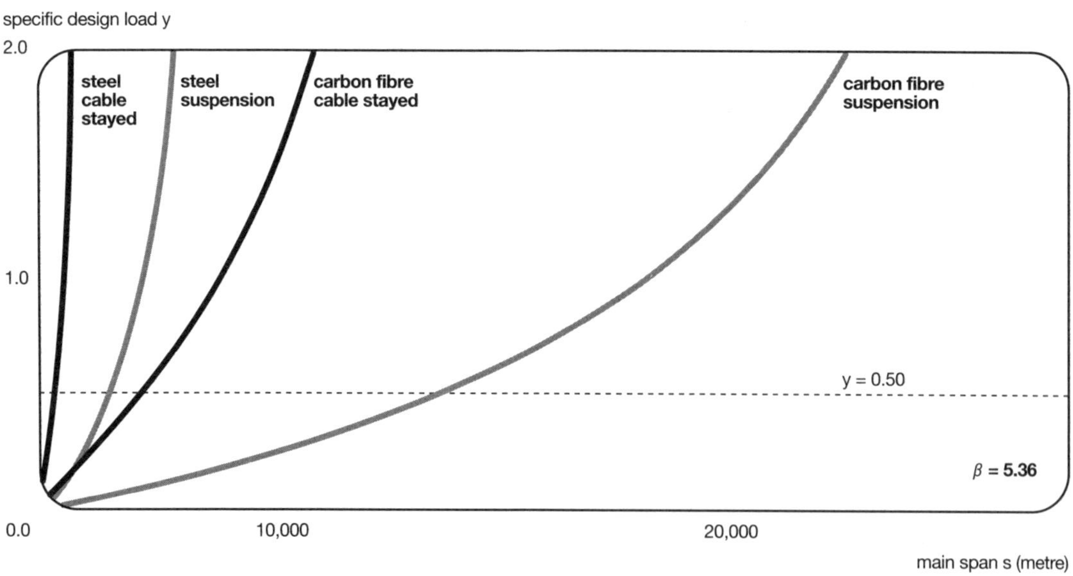

GRAVITY'S CENTRIFUGAL NEPHEW. Centrifugal forces in small rotating components, that could fulfil their duty in aircraft, act on fibre structural elements almost in the same way as global gravity forces act on big mechanical systems. For a free hanging vertical prismatic rod the maximum length is limited by the maximum stress in the fixing area.
This classical gravity case is mainly of interest in distinguishing the efficiency of different kinds of cable materials.
By rotating a cable or rod around a hub the allowable stress in the fixture area limits the maximum circumferential speed according to a comparable formula.
For these small scale structures 'gravity' is surpassed by 'centrifugal' load. In flywheel design the centrifugal force (density, rpm and radius) is not just a limitation. It can even serve to manufacture perfectly shaped 3-D tensoid structures. A dry wound tubular fibre structure between two annular fixtures can be centrifuged to transform and distribute the individual fibres in natural curves similar to the aforementioned catenaries. By proper anchoring, the distribution of fibres in the radial and tangential direction can be optimized for typical flywheel functionality, maximum energy storage per unit mass.
In cases where rotation is used for forming and manufacturing, the distribution of fibres and 'freezing' of the resulting structure by impregnating it with a resin and curing it, is quite a challenge for the development of lightweight high velocity flywheels.

$$l \leq \frac{\bar{\sigma}}{\rho g}$$

The maximum length of a prismatic rot limited by the maximum stress in the fixing area

$$\omega R \leq \left\{ \frac{2\,\bar{\sigma}}{\rho} \right\}^{1/2}$$

The maximal circumferential speed

$$e_m \leq K \cdot \frac{\bar{\sigma}}{\rho}$$

Max. energy stored per unit mass

Where K expresses the efficiency of the design

$K = 1/3$

just radial reinforments,a spoke or brush wheel

$\bar{\sigma} = \bar{\sigma}_{radial}$

$K = 1/2$

just tangentials, a rim or annular wheel

$\bar{\sigma} = \bar{\sigma}_{tangential}$

$K = 1$

optimal design, the Stodola wheel (isotropic)

$\bar{\sigma} = \bar{\sigma}_{radial} = \bar{\sigma}_{tangential}$

GRAVITY'S HYDROSTATIC NIECE. When internal pressure (p) exceeds the force of gravity as happens in a thin walled geodetically wound pressure vessel, a special case of fibre load occurs. Not only is it pure tension, it is also the same throughout the total length of the wound strand of fibres. Here we have arrived in the world of 'isotensoid' filament wound structures. They are dimensioned and stressed according to the 'netting theory'. The binding resin theoretically is irrelevant, due to its low Young's modulus in comparison to that of the applied fibres. The optimum vessel design, with or without cylindrical section, can be quantified by the maximum volume versus the structural weight.

$$\frac{V}{W} = \frac{\bar{\sigma}}{3\rho p}$$

When all tensoid structures are judged for structural efficiency, which means maximum performance per unit weight (or just minimum weight, or quantity of fibre/cable material per unit performance), the governing equations show the same specific mechanical property. For strength s/r and for structural stiffness Young's modulus E/r. Without any prejudgement of the class of cables or class of strands, the following tables A&B can be used for selecting materials, bearing in mind that some synthetic polymer fibres cannot withstand ultraviolet radiation (aramid fibres), some are very sensitive to humidity (natural fibres) and some 'creep' because they just can't stand high constant loading (E-glass).

B. Elastic properties of some fibres and steel compared

material		density ρ [10^3 kg/m^3]	young's modulus E [10^9 N/m^2]	yield stress σ [10^6 N/m^2]	specific modulus E/ρ [10^6 m^2/s^2]	specific strength σ/ρ [10^3 m^2/s^2]
steel	piano wire	7.8	210	3100	27	397
natural fibres	jute	1.46	10-25	400-800	12	411
	hemp	1.48	26-30	550-900	19	490
	sisal	1.33	46	700	35	526
synthetic organic fibres	HT-carbon T300	1.76	230	3530	131	2010
	HM-aramid	1.45	133	3500	78	2410
inorganic fibres	E-glass	2.58	73	3450	28	1340

note 1 properties of natural fibres are very variable, the figures shown are averages in tension and collected from many different publications.
note 2 yield stress is the lowest stress at which a material undergoes plastic deformation or failure.
note 3 the material values for fibres applied in unidirectional composites must be modified with the proper rules of mixtures (V_f).

Superfly.

Amateur airplane builder workshops

500
PH-COR
PH-COR
PH-COR

struct
c Weld

harris
3440
106
84
Matt Mid Stone
Gris Pierre Mat
Matt-Steingrau
BEAVER
85
88
Matt Deck Green
Vert Pont Mat
Vert Russe Mat
Matt-Russischgrün

McDONNEL
MiG-15 UTI
KP
PLASTIKOVY
MODEL
2513

F-15C EAGLE
4759
Revell

Swanenberg Heli Services bv
NIEUW!
2003
JUBILÄUMSFLUGT
17. und 18. Mai 2003
Flugplatz Oberschleißheim

BLACK METAL DESIGN. The common practice of replacing aluminium one-to-one with carbon reinforced epoxy resin, is called ‘black metal design’. If structural strength and stiffness were the only properties at hand, this approach could indeed lead to maximum weight reduction. Unfortunately for the conventional thinker, and indeed fortunately for the open-minded designer of composite structures, this is not the case. The current generation of ‘black metal’ designers ignores not only the weaknesses of current laminates, such as the risk of delamination at locations in free edges, but also the vast number of requirements a structure would meet as part of a larger system. For instance, a pressurized aircraft cabin has to possess strength, stiffness and resistance to all possible external influences. Additionally it should ensure that passengers have a chance to survive in severe climatic conditions and in the case of a crash.

Design requirements for pressurized fuselages

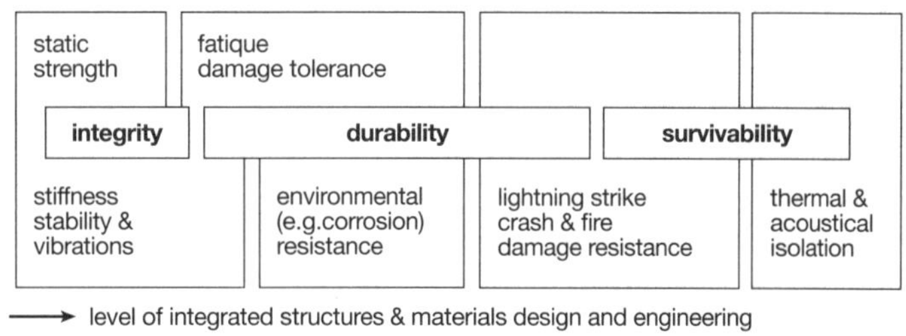

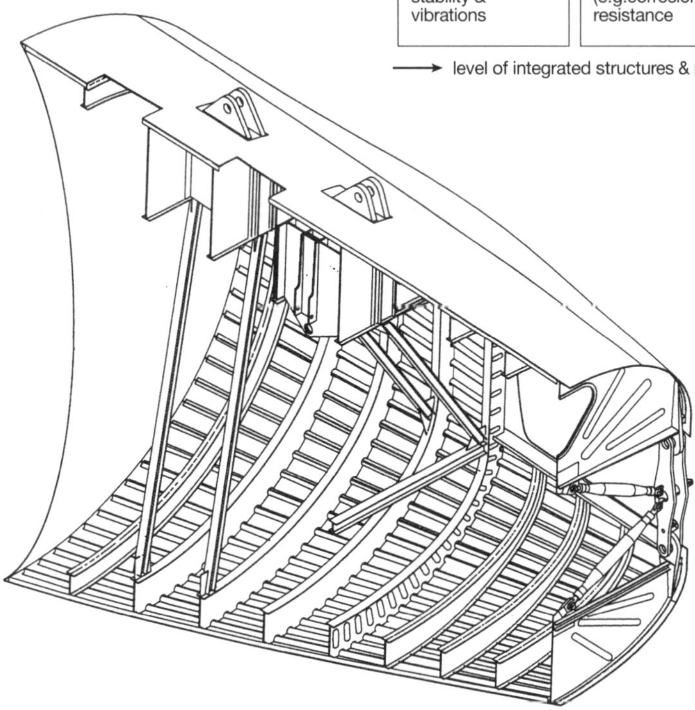

Structure efficiency of compression panels with different types of stringers

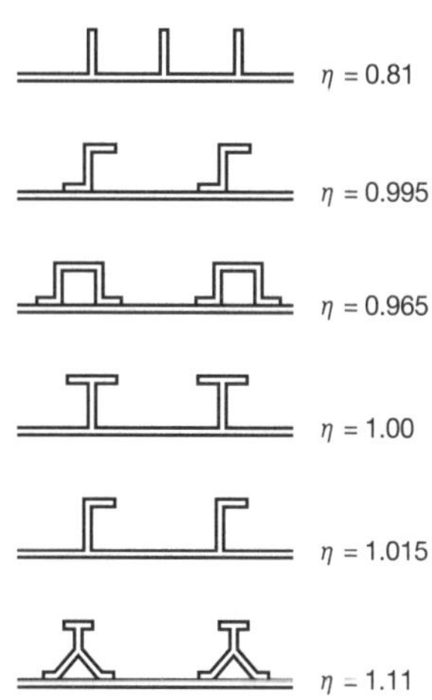

GRAVITY AND FRAME PITCH. The importance of small steps in the integral design approach is demonstrated by the diagram showing the influence of an elementary design parameter: the relationship between the choice of frame pitch, the distance between ribs and the resulting weight of a 10 metre long section of the Airbus A320 fuselage. The graph also shows the 'solution room' that emerges when one replaces the aluminium skin with fibre reinforced epoxy sheet according to the 'black metal' method.

The new free solution room appears between the two lines that constitute the structural lower limit for the two compared materials. They reflect the fact that below a certain frame pitch the weight of the fuselage remains constant and is determined by the permissible pure compression in the skin. For a larger frame pitch, the resulting weight is dependent on the skin thickness that determines the stability and resistance against buckling of the free skin panels (for the moment ignoring the presence of longitudinal stiffeners). The larger the frame pitch, the larger the required skin thickness to sustain buckling and because the load remains the same this leads to lower operational stresses. The lightest possible fuselage structure, a dream that still dominates the minds of the majority of aircraft developers and their bosses, is the one with the thinnest skin, ignoring all other requirements that a fuselage should satisfy. The design diagram shows that the lightest aluminium structure is about 30 per cent heavier than the lightest carbon epoxy version. It also shows that increasing the frame pitch – not related to seat pitch – from the traditional 20 inches or 500 mm (allegedly chosen because of the length of the horizontal ruler on a traditional drawing board) provides us with a domain where, at a slightly smaller weight reduction, the stresses in the skin are much lower, because of which fatigue and damage tolerance are no longer major factors. At the same time a weight and cost reduction is achieved by decreasing the number of frames, up to as much as 50 per cent. Naturally this has an effect on the way in which systems, floor panels and interior parts are attached to the fuselage, but many alternatives and even weight and cost reducing solutions are available. Notably, if the fuselage is transformed into a smooth sandwich structure, supports can be provided in the shape of lengthwise tubular elements, such as keel beams. They no longer have to be discrete elements that can only span the fuselage perpendicularly.

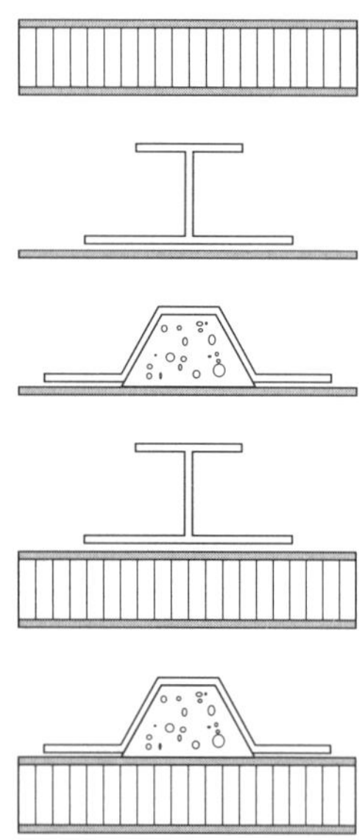

Boeing commissioned designs for different frame solutions but the space in between them was firmly established at 20 inches.

Fuselage weight tendency graph

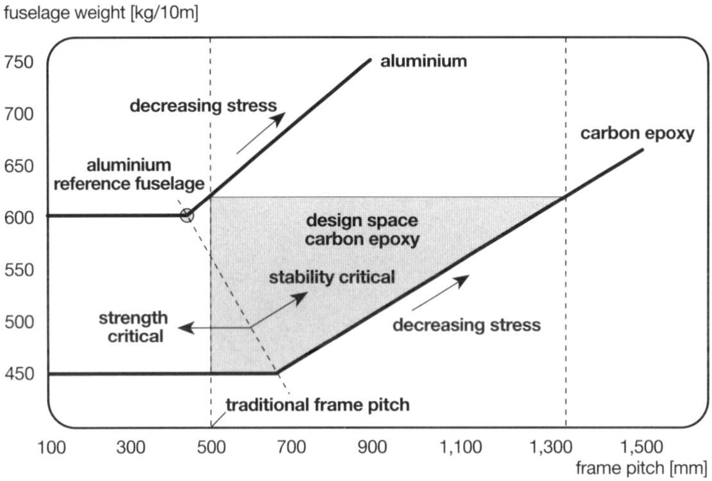

Influence of the frame and stiffener pitch on the TL, relative to the aluminium reference fuselage

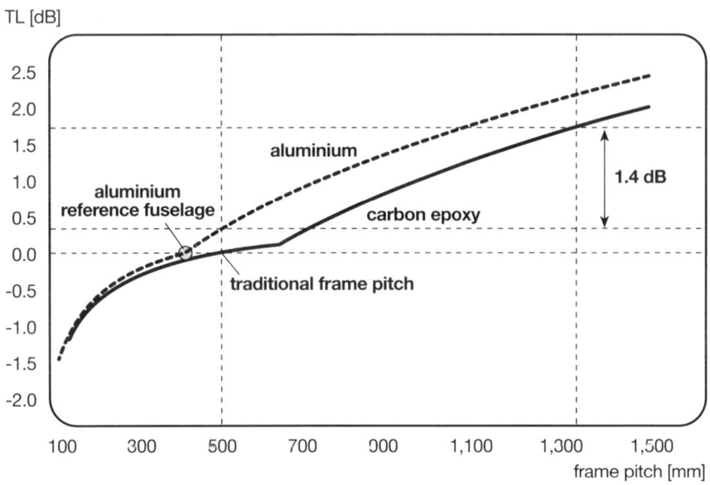

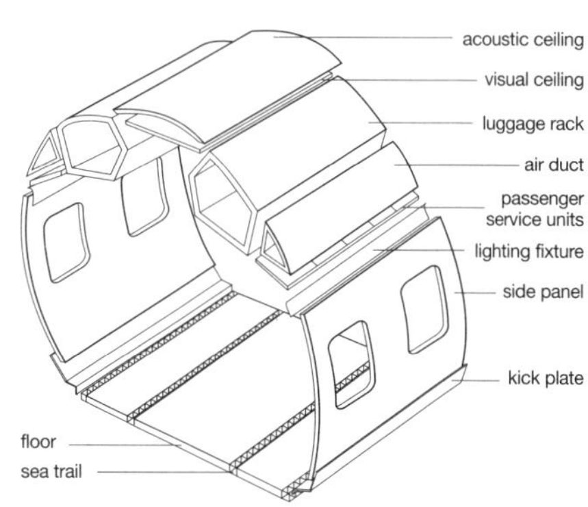

With black metal design one is able to demonstrate the benefits of letting go the traditional 500 millimetre frame pitch: fuselage weight decreases and sound dampening improves.

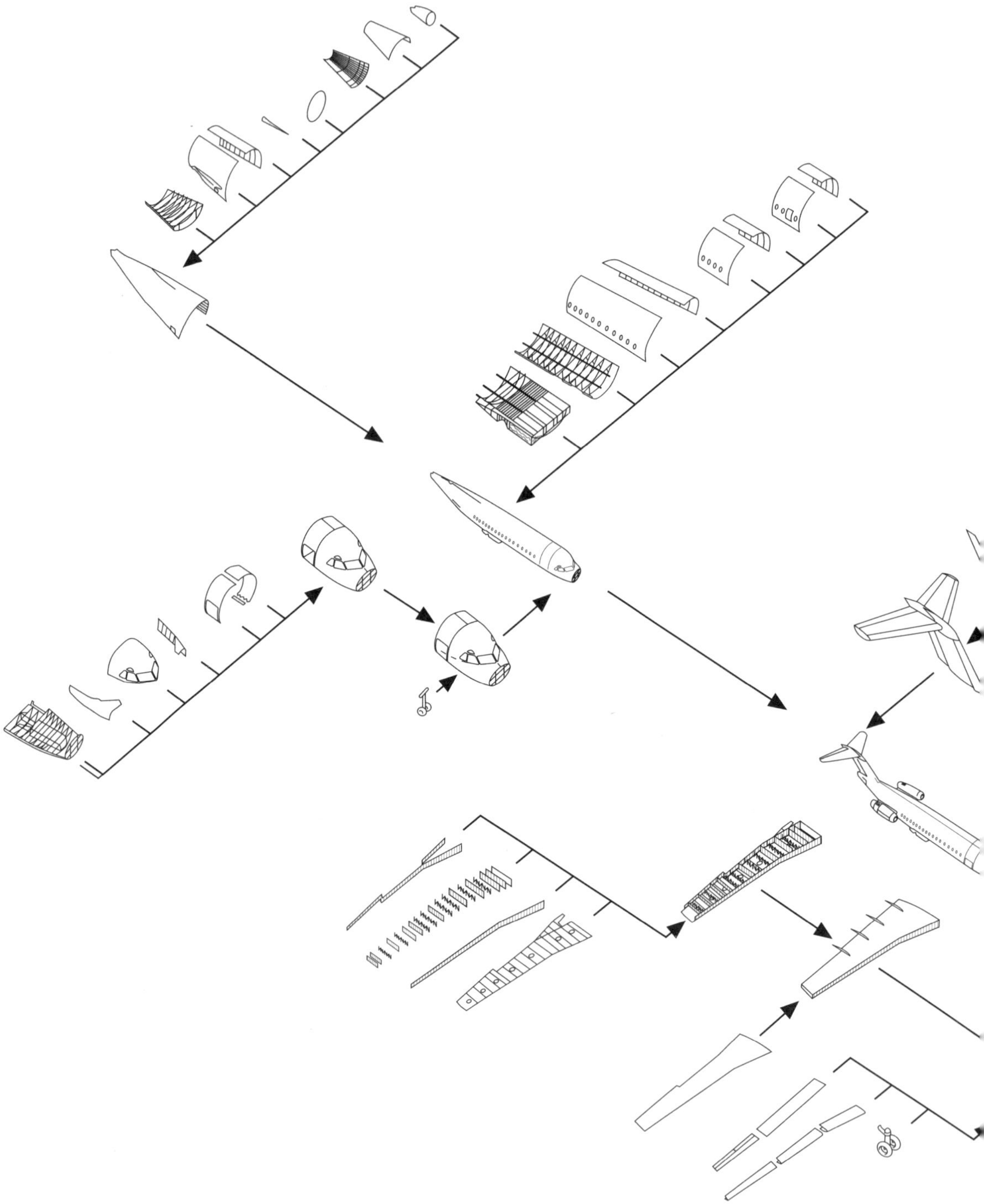

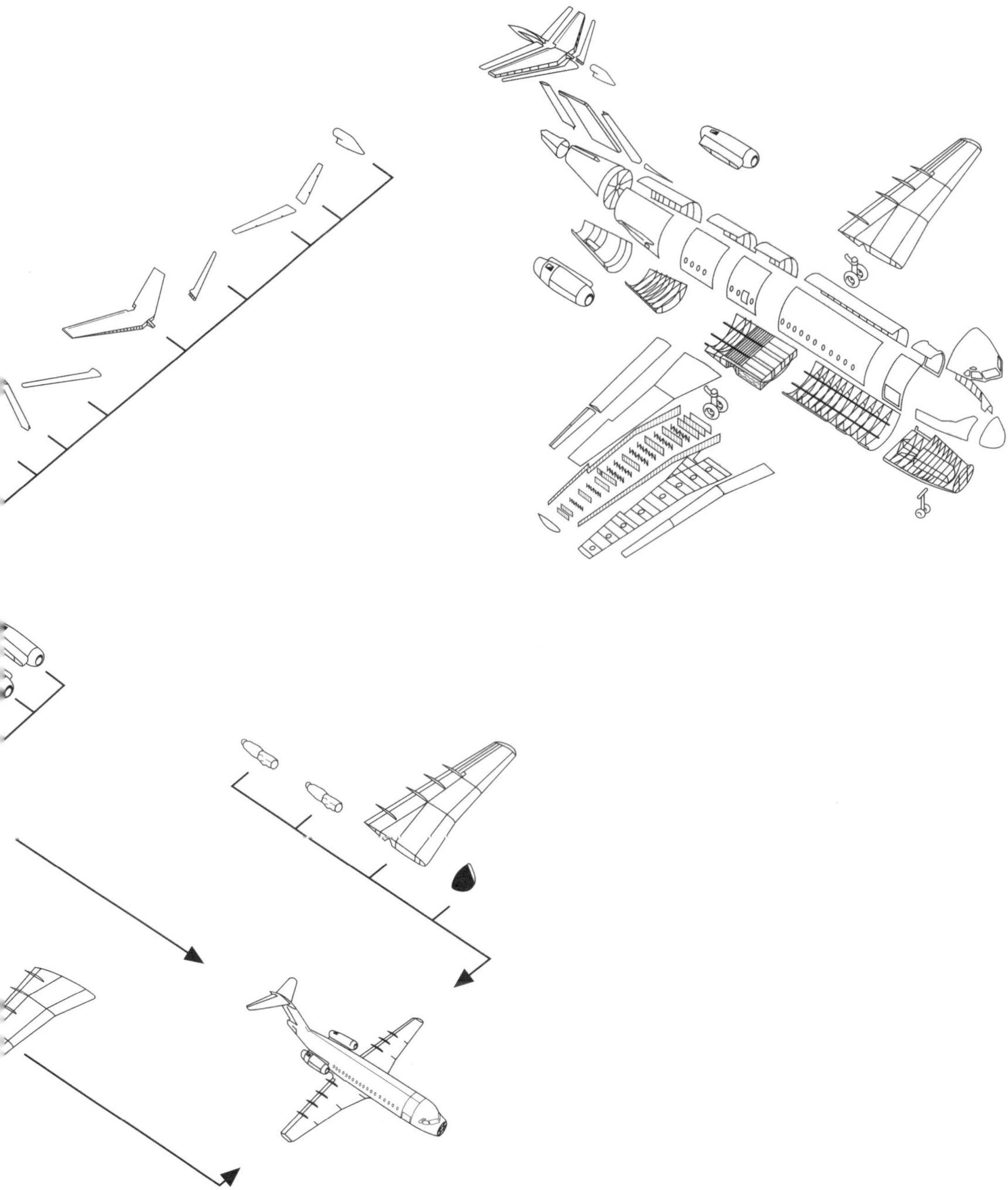

The cloth that dreams are made of

It is not easy to compete with polymer matrix composite structures against well-established metal solutions at the peak of their success. For a composite to replace an existing alternative it has to be both better and cheaper, and it has to have a culturally modern appeal too, as metal did in the 1930s. Composite materials do feature several significant advantages: they combine high stiffness and strength with low density, their internal composition can be controlled to the extent that they allow implementation of totally new functional and structural design concepts (like wings that can change shape according to air speed), and their resistance to corrosion is usually excellent. They are also attractive in terms of money: they can be used for the production of complex shapes, and raw materials can be inexpensive, especially with affordable glass fibre reinforcement. Apart from that, their use can generate significant savings in the area of service and maintenance. The growing market of composite consumer products, such as bicycles, proves that composites are considered up to date high-tech materials.

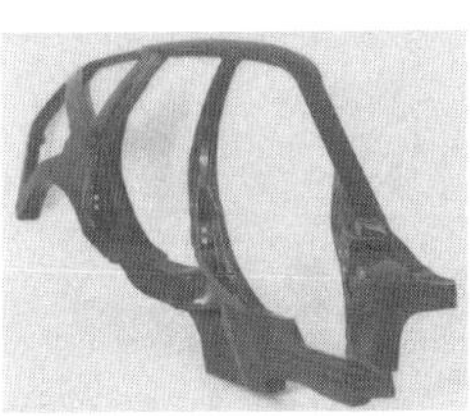

At a distance a textile reinforced polymer structure looks like metal. Touch it and knock on it and it even feels and sounds like metal.

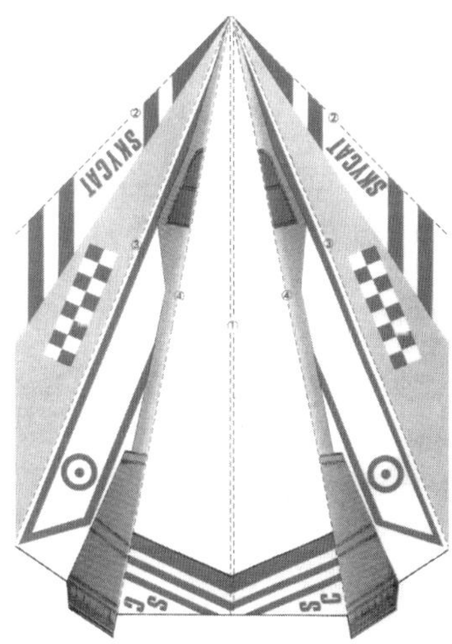

Designing textiles requires a different mindset.

Nonetheless, substituting composites for metal doesn't solely depend on what you get in the end. Designing with composites is often quite a challenge, because it implies inventing the material structure itself. In addition composite properties

usually vary in different directions. They are 'anisotropic'. The wide array of possible constituent materials – there are many kinds of polymers and fibres – and fibre architectures, plus the range of applicable manufacturing techniques, further increases the relative difficulty in choosing. Economic advantage is difficult to establish beforehand, because there is little experience and it depends on the adopted viewpoint, in particular on whether the sale price of the component is the sole consideration or whether the entire life cycle of the component is analysed, maintenance included. Finally, manufacturing is a challenge because the production of faultless material at low cost still can be quite difficult.
Despite these challenges, composites are now part of everyday life, and have entered nearly all major industrial sectors, including aerospace, ground transport, packaging, the sports industry and civil engineering.

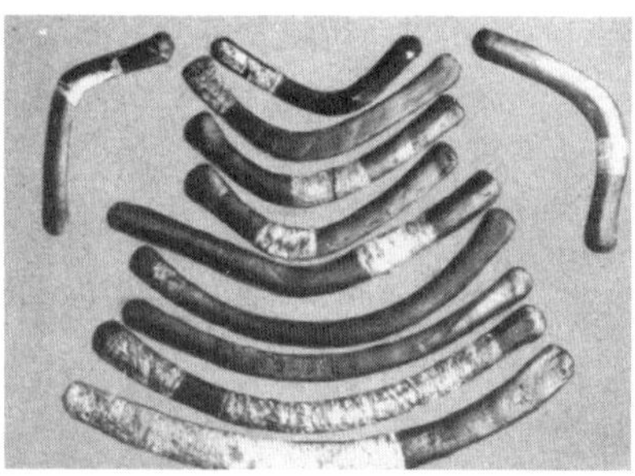

Ancient weapons were sometimes made with astonishing precision. The working principle of the boomerang, of which several kinds exist, wasn't discovered until the 1980s.

By marrying high performance fibres with a polymeric matrix, both the low mechanical properties of unreinforced polymers and the poor compressive strength of unbounded fibres essentially disappear, as does the inherent brittleness and weakness of each ingredient. In ancient times and also in current military applications the main rationale for implementing composites was, and still is, performance improvement. Nowadays, the use of frequently expensive materials can only be justified in designs by added value in performance or by cheaper manufacturing and maintenance. Therefore the successful design and development of

FASHIONABLE FIBRES

Fibre reinforced polymers have been in vogue for quite a long time already, although they have been popular under different names, mainly in the world of sports and expensive cars. The first composite that made it to the general public as a myth of new design opportunities was 'fibreglass'. It started out as a superior material for sports equipment in the 1950s. Chevrolet built the first fibreglass production Corvette sports car in 1953.

Carbon fibre reinforced polymers have reached that same kind of popularity now. They are applied in sports equipment, Formula 1 racing cars, racing bicycles but also comfortable and lightweight consumer folding bikes. Their appearance is considered 'cool' – so much so that some consumer products get covered with a print that makes them look as if they were made with carbon fibres, which they're not.

composite engineering structures benefits from maximum performance combined with a minimum use of materials, energy and cost over the entire lifespan.

The B2 stealth bomber is the first truly operational flying wing and also the first full composite aircraft.

Compared with traditional metal designs, modern well-designed and well-manufactured composite structures offer higher mass-specific stiffness and strength as well as better resistance to fatigue and environmental degradation. Aircraft propeller blades and helicopter and windmill rotor blades have already proved their quality. When it comes to crash or impact energy absorption, composite solutions can become vastly superior. Formula 1 racing cars, train fronts and armour all bear witness to this. Where damage resistance or damage tolerance are important, laminated polymer matrix composites become superior to most steel and aluminium designs without weight penalties, provided they are correctly designed and manufactured. Wings and fuselages of the latest military fighters and business aircraft illustrate this fact. By far the biggest potential benefit of polymeric composites is their ability to improve structural performance both per unit weight and per unit cost, thanks to the fact that their 'shapeability' is virtually limitless. Moreover they allow various mechanical, physical and chemical properties to integrate in one and the same concept. Examples of this are abundant and include the B2 stealth bomber, sailplanes, pressure vessels, temperature conditioned transport, packaging, bridges and canal locks. The fact that functionality can be defined all the way from the smallest fibre and the tiniest lump of polymer to the entire structure really is the unique selling point of composites.

ANCIENT COMPOSITES

The principle of combining different materials to get a synergetic artefact dates back thousands of years. The application depended on what people had at their disposal and on what they needed. Wattle and daub or straw reinforced clay became important building materials for permanent protective settlements. The earliest laminated composite tools, such as lightweight bows and chariots, consisted of glued layered structures of wood, hide, bone, sinew and horn. The use and combination of those natural materials, polymeric and fibrous, based on chains of protein, cellulose, collagen and keratin was a development executed by prehistoric craftsmen. They didn't know all these abstract terms, but they did already understand that, by a proper marriage of different materials, they could exploit their typical beneficial properties, while the drawbacks of the individual components often disappeared when they were combined into a composite material.

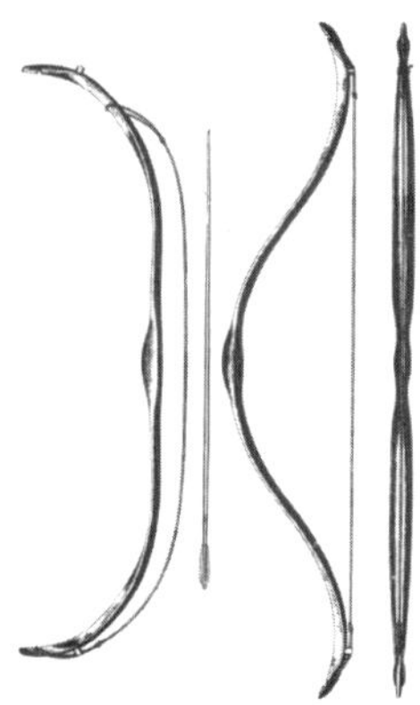

Although nowadays there are various production techniques for every product size, series length and quality, for the envisaged large airplane structures the focus will be on low pressure resin injection techniques, in which the core of the construction consists of fibre structures produced through very familiar methods such as winding, weaving, braiding and stitching.

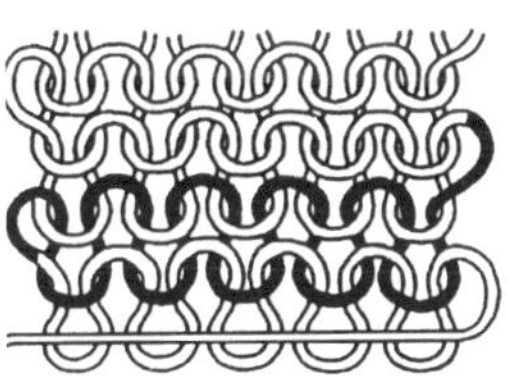

We've been wearing stuff like that for ages. In the case of airplane structures they are invariably made from flat filaments of mainly E-glass or HT-carbon that can be industrially woven. The resulting textiles can be assembled in a mould into supporting cores of three-dimensional shell structures, from sandwich shells with two skins and a foam core to massive shells, locally reinforced with stiffeners. Depending on their shape these textile pre-forms can be impregnated before, during or after assembly. This can be done with thermoplastic or thermoset polymers, or even with rubbers, for pliable textile structures can also be made into flexible multilayer structures by embedding them in a flexible material for special clothing, tent-shaped or inflatable structures equipped with all possible barrier functions, and data and energy transport facilities. Depending on the product size and the produced number, there are many techniques already available – and some are still being developed – for filling a mould or impregnating fibre structures.

For very large constructions it remains imperative to operate at low pressure, and to use a simple oven if curing is needed. For wound pressure vessels and columns the plastic mould could also end up as an air- and fluid-tight layer

PAMPER THE PAYLOAD

Economy class is extending its territory at the cost of more expensive travel seating. An airplane chair determines passenger comfort. The British design firm Thomson Solutions is proposing different chair arrangements that provide both more space and more efficiency. The chairs are arranged in such a way that the arm support of one chair overlaps with the leg space of the next. The one is placed obliquely behind the next. The result: two or three inches more width and 15 per cent more capacity. For a blended wing configuration this kind of arrangement would perfectly fit the shape of the plane.

The first fully composite passenger aircraft is scheduled to fly in 2008. It will be the Boeing 7E7 'Dreamliner'. The configuration is traditional, but the main structure will consist of carbon reinforced epoxy resin and some parts of the wing will be made of titanium graphite composite. The textile resin combination provides greater freedom in shape, thereby improving aerodynamic quality. It is remarkable that the manufacturer chooses to employ the strength of composite materials to improve climatic conditions in the passenger cabin. The air pressure will be higher inside. It will be maintained at the level of 1800 metres instead of the now common 2400 metres. Because the material will have far better insulating properties than aluminium, condensation of water can be reduced, resulting in a more pleasant atmospheric humidity. In addition the 'Dreamliner' will have the biggest windows ever made in a jet airliner: 48x28 centimetres. The amount of light coming through can be electronically controlled.

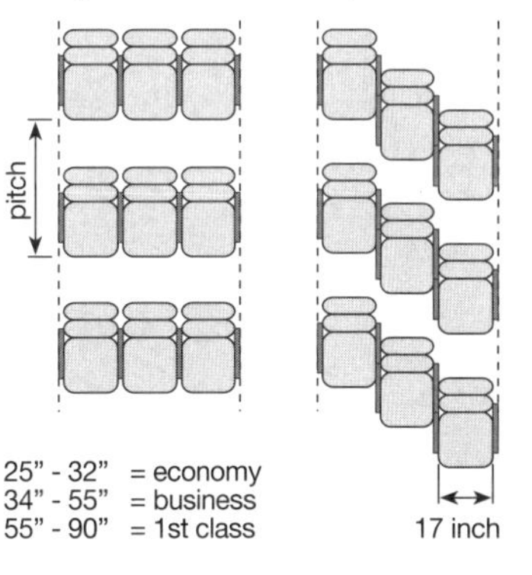

Textiles don't always need to be impregnated.

on the moulded part. Since it is feasible to arrive at the final product through a minimum of materials and elements that could be considered 'the ingredients', and a limited number of steps in the production process from fibre to textile structure to product as 'the cooking', this could yield substantial logistic and financial advantages for a manufacturer. Almost any product can be manufactured.

Table 3a. Engineering constants of different yarns

		engineering constants		carbon fibre yarns	glass fibre yarns	alu fibre yarn	steel fibre yarn
				existing T 300	existing E-type	virtual 7075	virtual piano wire
Material properties	density	ρ	$[10^3\ kg/m^3]$	1.76	2.58	2.70	7.80
	young's modulus	E	$[10^9\ N/m^2]$	230	73	71	210
	yield stress	σ_y	$[10^6\ N/m^2]$	3530	3450	420	3100

In order to select the appropriate fibre type for the envisaged light shell structures, one has to think in terms of design. Here we limit ourselves to rigid structures for aerospace engineering, but the same principles are valid for applications like accelerating machine parts, low-speed trains, sporting equipment, oil pipes and civil superstructures such as suspension bridges and skyscrapers. For various complex structural elements, most of which can really only be produced as composite structures, the performance of glass and carbon fibres as well as high-quality aluminium and steel fibres is shown in table 3a. Let us conveniently assume that the fibre volume and architecture as well as the eventual fibre-matrix adhesion are always the same. The numbers indicate the structural efficiency of the fibre reinforced structures, meaning the capability of carrying load per unit mass, scaled to the least efficient material (1.0). Inverting these indices gives the potential weight reduction per component. Then we can conclude that the carbon fibre will in all cases lead to the lightest structure. 102

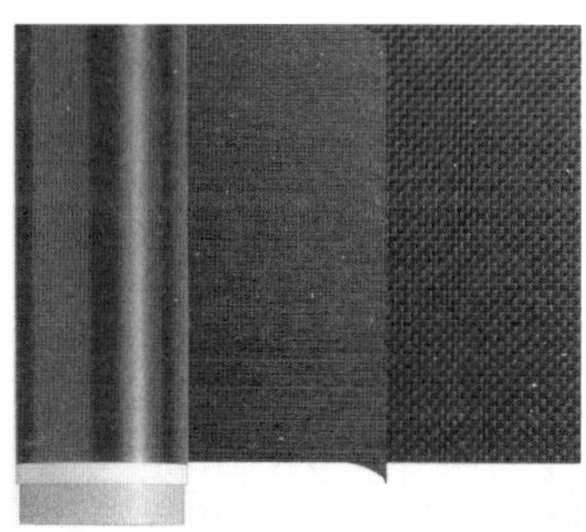

Textiles can be made from different fibres and in different patterns. The possibilities are limitless.

Particularly in the case of the wound pressure vessels the differences are substantial. For curved sheets and shells, which are most likely to buckle, we note that cheap E-glass and high-grade 7075 aluminium are close competitors, but not when it comes to material costs and resistance to corrosion. The same applies to curved sandwich shells, two loadbearing skins with a very light foam core between them that also enhances thermal and acoustic insulation. Especially surprising is the large advantage of carbon fibres over high-grade steel in pressure vessels. For an equal load, we can reduce the weight by about 80 per cent if we simply replace one material directly with another. For curved beams (wildly curving if necessary) that are critical with respect to bending stiffness the figures are once again convincing. Carbon can withstand bending 4.7 times better or yields, under equal loads,

PRICKLY NOSES

In the Second World War both the Allied forces and the Germans had radar built into the noses of some airplane types, but the Allied forces had composites. The Germans had to build an entire forest of 'Liechtenstein' antennae on the outside of the metal nose of, among other planes, the Dornier Do217-N2 night fighter. It caused quite a lot of drag, brought down maximum speed and increased fuel consumption. The Mosquito, one of the most successful planes ever, can be considered a composite plane predecessor, and it had a composite nose cone to cover the antenna and minimize drag.

Table 3b. Structure efficiency of different yarns

Structure efficiency versus the lowest listed value (1.0)	based on imaginary yarn properties of equal appearance	dominant parameter	carbon fibre yarns	glass fibre yarns	alu fibre yarn	steel fibre yarn
SOLID SHELLS						
Compression	buckling critical	$[E^{1/3}/\rho]$ vs. $[E^{1/3}/\rho]_{min}$	4.5	2.1	2.0	1.0
CURVED BEAMS	buckling critical or	$[E^{1/2}/\rho]$ vs.	4.7	1.8	1.7	1.0
Compression	stiffness critical	$[E^{1/2}/\rho]_{min}$	4.1	4.1	1.0	3.9
Bending	strength critical					
Bending		$[\sigma_f^{2/3}/\rho]$ vs. $[\sigma_f^{2/3}/\rho]_{min}$				
PRESSURE VESSELS	strength critical, isotensoid wound	$[\sigma_f/\rho]$ vs. $[\sigma_f/\rho]_{min}$	12.9	8.6	1.0	2.5
SANDWICH SHELLS						
Bending	stiffness critical	$[E/\rho]$ vs. $[E/\rho]_{min}$	4.8	1.0	1.0	1.0

again a weight reduction of about 80 per cent in comparison with the steel version. When the bending strength is most important, however, the glass and steel fibres surprise

100

Fashion designers have all kinds of textile working techniques at their disposal, like draping and moulage. Composite engineers have only just begun to explore these opportunities.
The direction of pleats depends on the making.
Freedom of movement depends on pleats.

us yet again, especially when we consider the damage resistance and 'residual strength' (the part of the steel stress-strain curve that corresponds to plastic deformation). For in that case steel and cheap glass fibres are very strong competitors of carbon fibres.

Again, to avoid confusion, this only applies to these types of materials and the products that can be made from them. If metal fibres can enter into a perfect fusion with the metal matrix, let us say they become metal sheet, then the numbers become completely different since the properties for sheet apply in all directions. This is of course not the case with high-quality fibres fixed in a mechanically inferior polymer matrix. For a simple composite sheet, with equal quantities of fibres in two perpendicular directions and a fibre volume percentage of 50 per cent, the engineering parameters E and σ_y in both directions will be reduced by about 75 per cent. Moreover the specific weight, especially of metals, will change according to the law of mixtures (table 3b). Nevertheless, the same comparisons, but now based on sheet properties for the same structural elements (flat or single-curved), once again declare carbon reinforced epoxy sheet materials the winner.

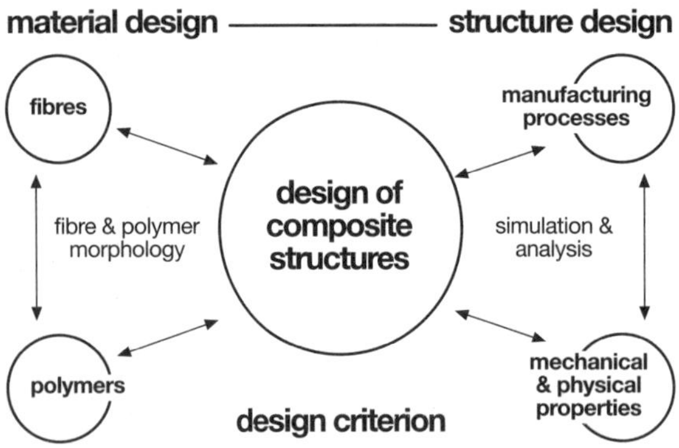

The design of a composite structure includes the design of the material and its local variations, as well as the manufacturing process.

The direction, which further development will take depends on structural quality, but at least as much on the general economical tendency to shift from labour intensive to labour

SMART PROMISES

Many developments are going on in the two basic elements that constitute composites: fabrics and matrices. To start with, the former, consisting of fibres and textiles these days can be coated, even on nanoscale, to provide them with strange new properties: they can clean themselves to a certain extent, thereby reducing use of water emulgators and energy. Textiles can also act as chemical separators of various substances, or as catalysts. They can be watertight and at the same time allow gases to go through. Goretex, originally developed for medical applications, is good at that. Furthermore textile texture can be designed three-dimensionally in such a way that air or viscous drag is reduced, as shown in sports costumes for skating and swimming. Metal fibres can be integrated that transform through heat (memory metals), or that conduct electricity, which also happens to be a property of some plastics. This property is applicable in the production of these materials themselves. Some other polymers possess piezoelectric powers: they transform through current and produce current when transformed. Still others produce light. In the meantime electronic components such as memory chips and transponders are shrinking to become small enough for integration into virtually any material. As a result composites can be designed to be smart. They can then sense environmental conditions and react to them, through integrated intelligence. Still it will be quite a while before synthetic structures are as clever as even the smallest bug.

extensive technology. More than a century ago the rapid introduction of electricity evolved from industrial capabilities to transforming natural fibres and fillers, phenolic polymers and natural rubber into insulating structures for cables, switches and electrical equipment. Automotive transport was revolutionized by the large-scale production of low-noise, lightweight inflatable tyres, thereby making cars less dependent on road quality. Although labour was inexpensive by today's standards, a successful breakthrough of these applications relied on industrial techniques capable of keeping up with mass production. In the end, nothing determines the price-performance ratio of commercial products as much as the cost of manufacturing. Composite helmets and LPG tanks are recent examples of that phenomenon. Once new materials and products have been accepted, the performance per unit weight becomes more important and a process of optimization begins. Decreasing structural 'dead' weight is part of that. It is of course beneficial for performance improvement and can be applied in packaging, transportation and in large structures whose function is limited by gravity, such as tall bridges and buildings. In addition, minimizing the quantity of material leads to reduction of material cost, production time and consequently of the cost of manufacturing. Nowadays, the success of polymer matrix composites can increasingly be measured by the accessibility and reproducibility of manufacturing techniques.
Factors determining the implementation of polymer matrix composites in engineering applications include the intrinsic performance of the materials and the economic advantages that can be gained through their use. These factors, in turn, are strongly dependent on the somewhat specific nature of designing with composites and also on the increasingly important role played by materials, processes and part manufacturing procedures. Given the sustained rate of progress in these areas, composite materials no doubt will continue growing in engineering importance for many years to come.

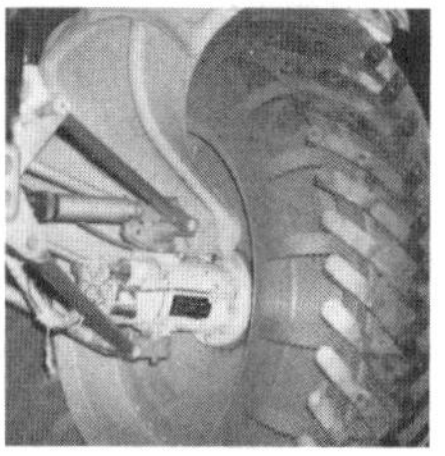

Top: Early, medium (antarctic vehicle) and recent (moon cart) interpretation of built-in elasticity demonstrate the artisan use of rubber, air and woven steel and aluminium respectively.
Bottom: Mass produced rubber tire

LITTLE FIBRES. The pioneering period for processes like injection moulding, flow pressing and form pressing of short-fibre reinforced polymers relied on thermosetting polymers ranging from phenolics, polyesters and vinyl esters to epoxies. In the case of injection moulding, these short fibres were less than about 10 mm long, usually glass and sometimes carbon, and in matched-die pressing they were longer (less than about 100 mm, usually glass). Gradually thermo set polymers are now being replaced by more expensive but technically equivalent or better thermoplastics. The main reason for this is the cost reduction due to shorter cycle times, which is crucial for mass production and recyclability. Reinforcement is limited with respect to fibre length, fibre content by volume (below 35 per cent) and orientation control. Injection moulded products are generally small, but they can be complex in shape and produced with very close tolerances. Camera housings are produced in this way.
Match-die flow moulding of larger plate and shell structures has also gained in importance. Stacked prefabricated sheets of fibre reinforced thermo set polymers (glass-polyester, sheet moulding compound or SMC) or thermoplastic polymers (polypropylene, glass mat reinforced thermoplastics or GMT) are brought into shape by pressing at the right temperature. The length (shorter

Production of carbon fibre

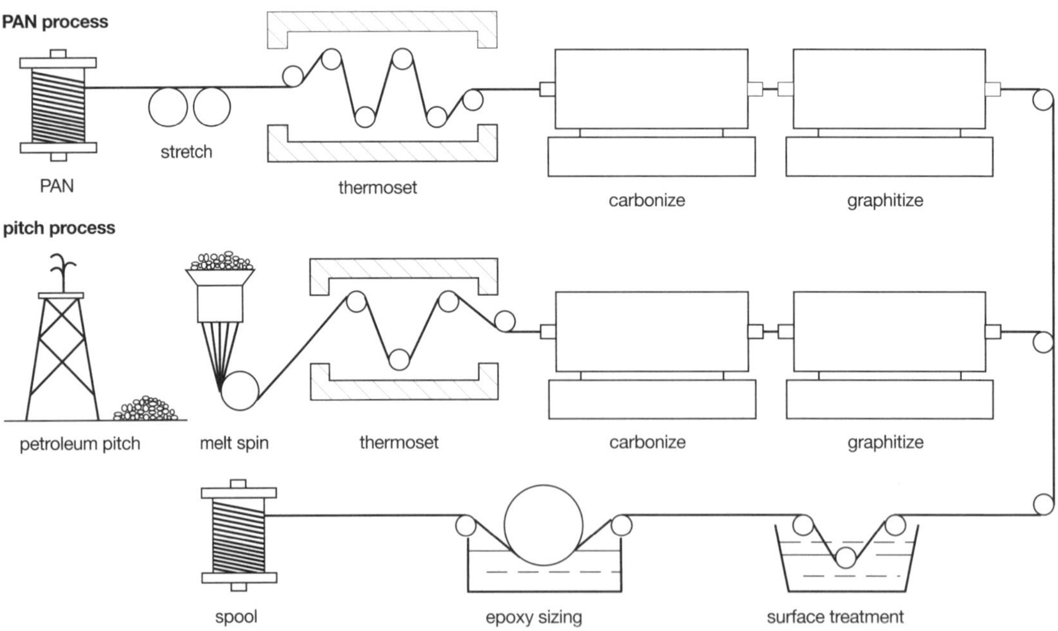

than 100 mm) and volume percentages (below 45 per cent) of the reinforcing fibres (typically chopped E-glass) are greater compared to injection-moulded composites. This results in stronger reinforcement and higher composite in-plane stiffness and strength. The control of fibre orientation is similar to that in injection moulding and is based on keeping the fibre distribution as random and uniform as possible – minimum control you might say – mainly in the in-plane directions. Typical applications of products of this kind of processing include parts for trucks and cars as well as garden furniture.
For both techniques, heavy machines are indispensable. Rigid temperature controlled moulds are kept closed with a force sufficient to overcome the pressure of the process (for injection moulding up to 1000 bar and for pressing 100-200 bar) multiplied by the projected part area. This force thus tends to be rather high. To compensate for the ensuing investment costs – high forces require expensive equipment – large numbers of composite parts have to be sold. Therefore automotive components, housings for cameras, popular electrical and electronic equipment, furniture, leisure and sporting goods have become popular applications for this class of composites, typically replacing metal and wood. The short and discontinuous fibre reinforced plastics are responsible for the biggest share of successful applications of polymer matrix composite materials.

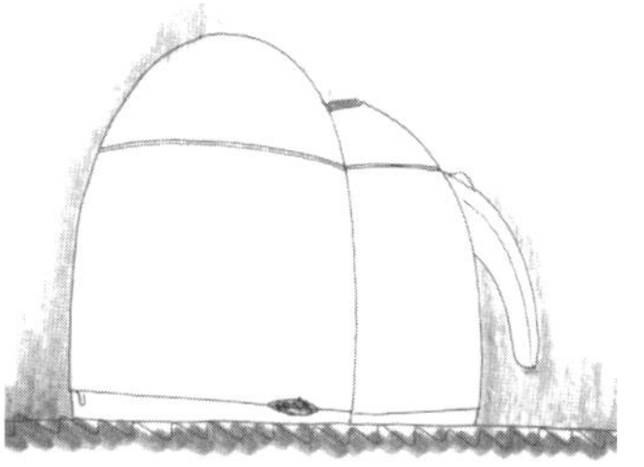

Fibreglass forming process

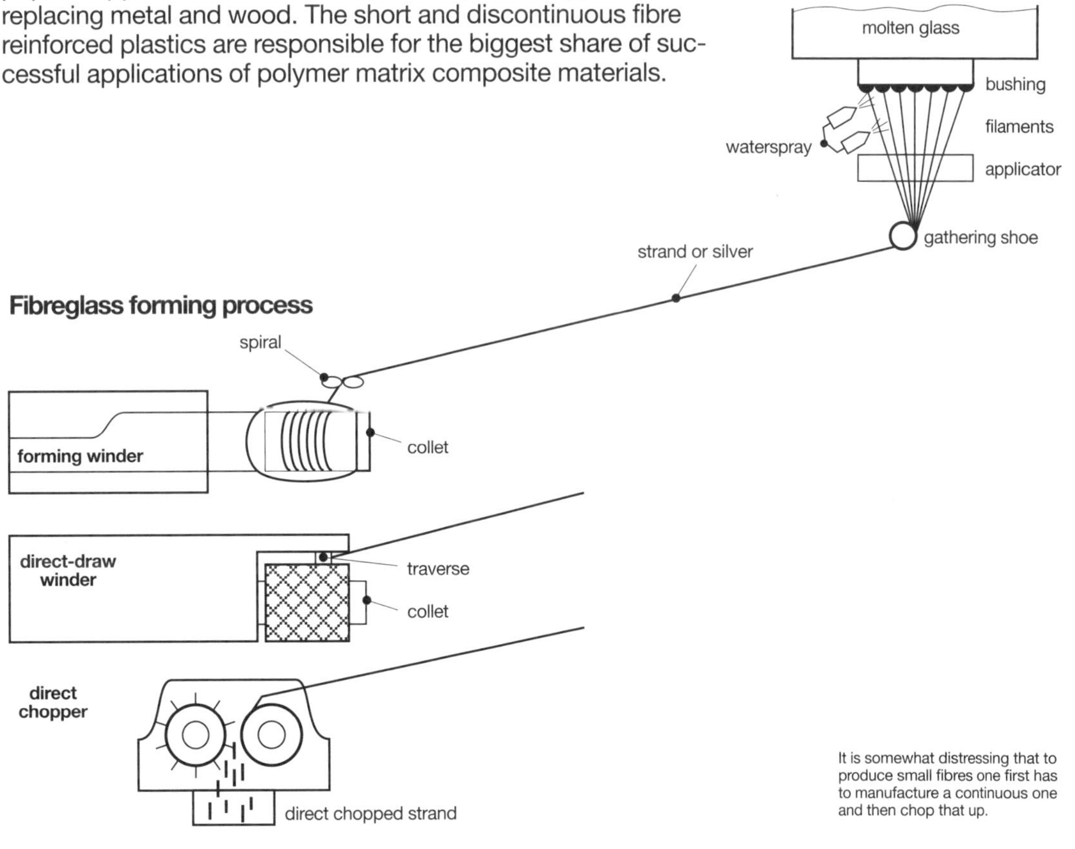

It is somewhat distressing that to produce small fibres one first has to manufacture a continuous one and then chop that up.

fibre content Vf [%]	orientation control [%]	fibre length L [mm]	applications	manufacturing processes
	0	**short fibres** $L_{crit} < L < 10$	**small detailed parts** housing and brackets	**thermosets** **thermoplastics** **injection moulding** **pressing**
25		**discontinuous fibres** $10 < L < 100$	**beams and shell structures**	**injection moulding** **(S)RIM** **resin infusion**
			car body parts	**pressing** **BMC/SMC** **GMT**
40	100	**endless fibres** $L \rightarrow \infty$	**large plate and shell structures**	**laminating**
			aerospace	**tapelaying**
			packaging/ pressure vessels	**filament winding**
			ship building/ civil engineering	**resin infusion**
			small plate and shell structures aerospace	**pressing** **diagraph forming**
70			**profiled beams** civil engineering	**pultrusions**

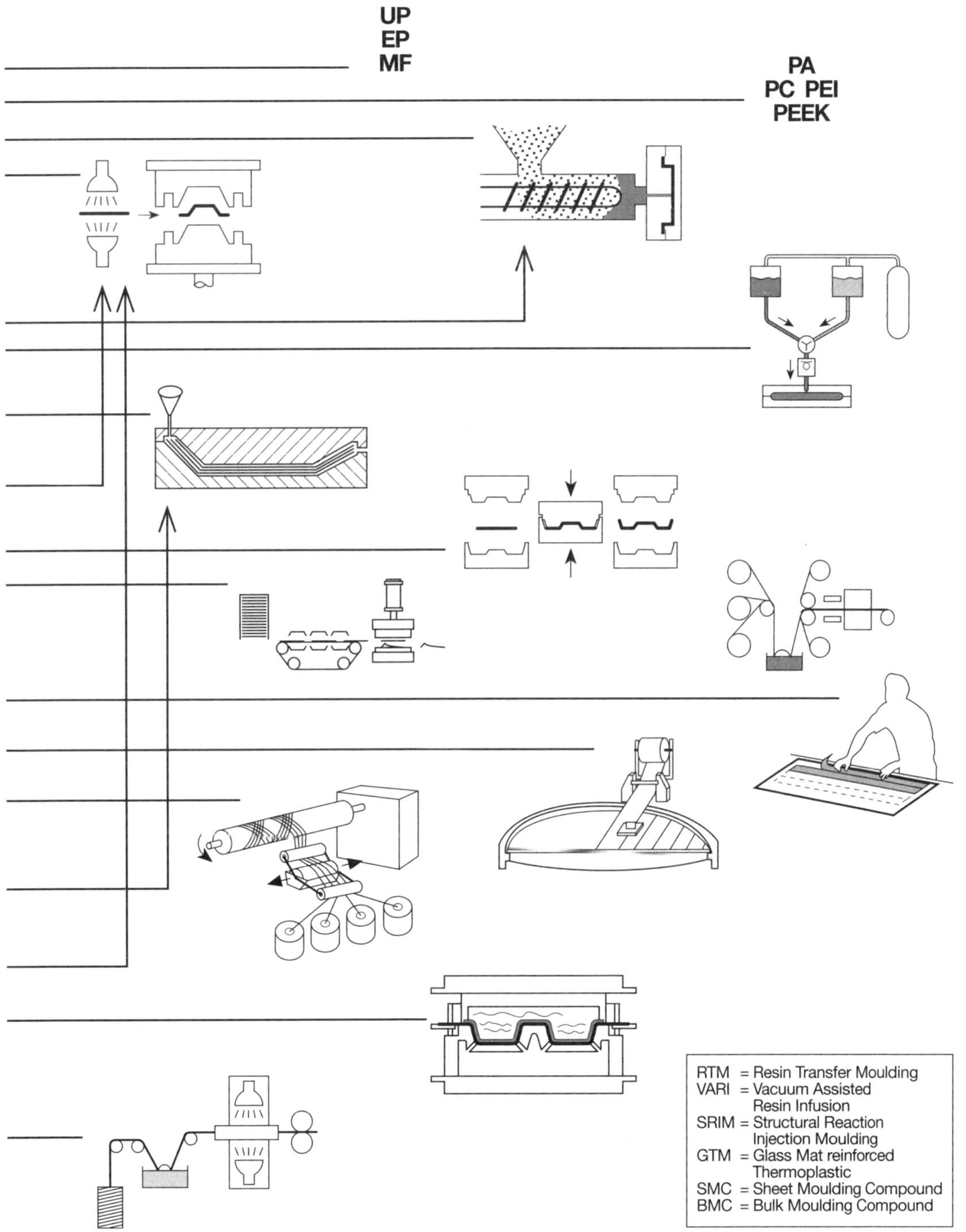
UP
EP
MF
PA
PC PEI
PEEK
RTM = Resin Transfer Moulding
VARI = Vacuum Assisted Resin Infusion
SRIM = Structural Reaction Injection Moulding
GTM = Glass Mat reinforced Thermoplastic
SMC = Sheet Moulding Compound
BMC = Bulk Moulding Compound

ENDLESS FIBRES. The number of applications of continuous fibre reinforced polymers has grown enormously since the last decade. Their domain generally consists of advanced capital-intensive thermo set-based materials and products. Except for fast (reciprocating) machine parts and expensive and fashionable leisure and sporting goods (alpine sports, car racing, angling, tennis, golf and cycling), most of the applications in continuous fibres are bigger in size and take the form of plate and shell structures. Big structures (low numbers, big volumes) are almost exclusively made from thermo set polymers such as phenolics, polyesters and, increasingly, epoxies. The reasons are both the relative low cost per unit weight and the low pressures and temperatures at which the structures can cure.
In aerospace applications, continuous carbon fibres are increasingly replacing glass fibres to reduce structural weight, unless low material cost, good performance in impact protection or specific electric or dielectric properties are required. For structural applications up to temperatures of 200 degrees centigrade, matrices usually consist of epoxy resins. For cabin interiors, the domain of stringent fire, smoke and toxic gas regulations, phenolics are very popular. For light and highly loaded structures a well-controlled and optimized fibre placement is essential. For this designers and manufacturers often still depend on expensive technologies. Laminating, for instance, can be done manually or by sophisticated automatic equipment. It depends on the number of parts to be produced and on the design. For fibre impregnation and cutting patches of unidirectional tapes or woven fabrics, human labour is gradually being replaced by cost-reducing accurate machinery. The most successful techniques to make composites are filament winding of pressure vessels and pultrusion of composite profiles applied in civil engineering: cables, bridges and electricity masts. Except for carbon fibres in pre-tensioned cables, glass fibres are the most popular in this class of relatively large applications. Aramid and carbon fibres are becoming strong competitors in pressure vessels, supported by the necessity of additional weight reduction and a reduction of winding cycle time to speed up production. Press forming is also gaining importance for continuous fibre reinforced products. Woven fabrics, for example can be draped into three-dimensional moulds in a resilient thermoplastic binder to keep the fabric 'frozen' in its shape. The result is a high quality preformed structure that can consequently be impregnated and cured to become the final product. Impregnation can be done in closed moulds by resin transfer moulding (RTM). Low pressure vacuum-assisted RTM, a cheaper alternative, can be executed in simple (female) moulds, with just a sheet of flexible film as counter mould. For additional pressure, relatively heavy and rigid matching dies are needed. Vacuum assisted RTM is becoming particularly popular for making very large composite

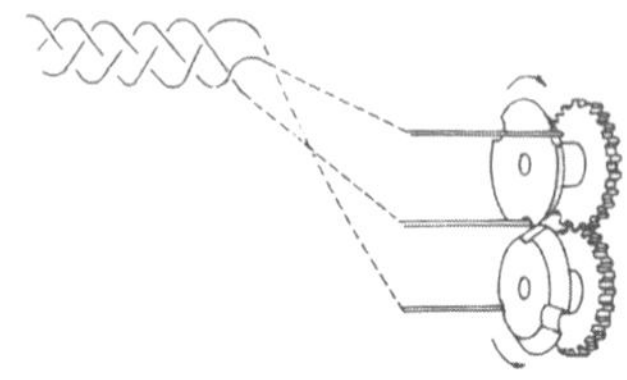

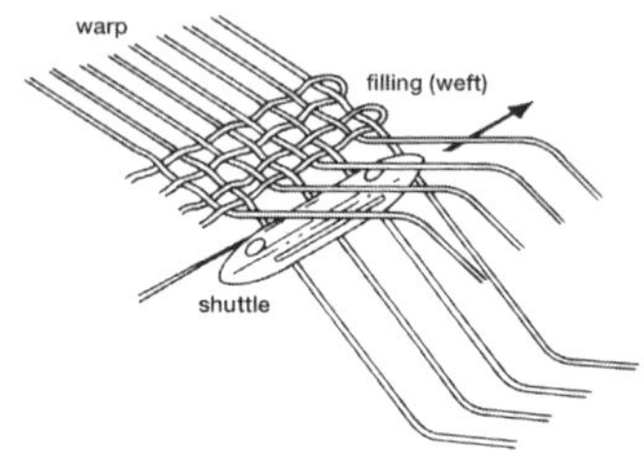

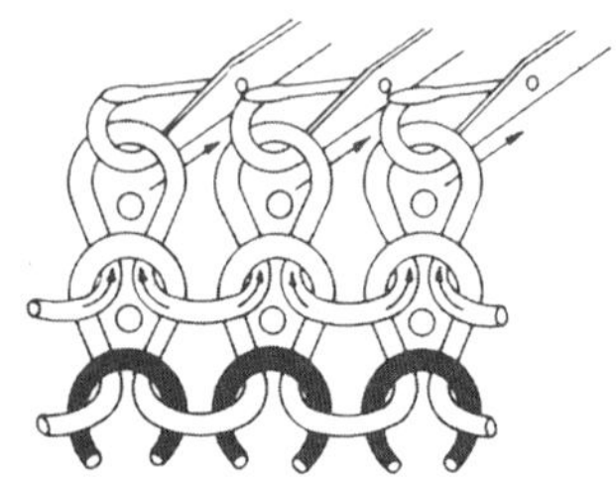

Right: The short and discontinuous fibre reinforced plastics are responsible for the biggest share of successful applications of polymer matrix composite materials.

structures, such as hulls and decks for ships or windmill rotor blades, with lengths of up to about 30 metres. 'Pressurized' RTM is applied in manufacturing small propellers and helicopter rotor blades. Here cost reduction, improvement of labour conditions, product and environmental quality are combined. For some applications, fabrics can be pre-impregnated with a thermoplastic polymer (powder, melt or solvent impregnation) and after heating, these 'prepregs' can be draped and consolidated in a mould. High fibre volume fractions and complete fibre orientation control are typical of this process and the speed is comparable to that of injection moulding and other industrial processes commonly used for short fibre reinforced polymers. The pressure levels however are very low: typically one bar or less for forming or draping and 10-40 bars for consolidation. Army helmets are produced in this way. When heating and cooling times are considered – a matter of seconds or minutes – it is clear that press forming is attractive not only because of the manufacturing speed but also for the lightweight equipment and minimum energy required.

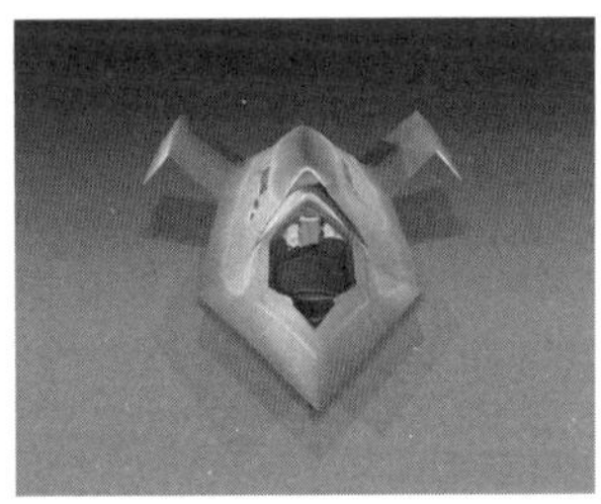

Resin infusion

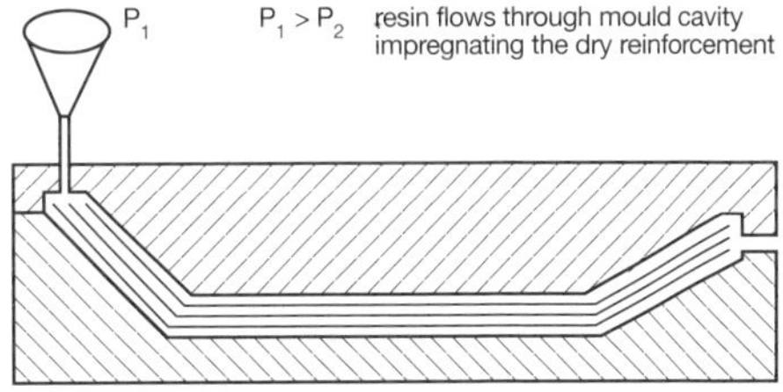

Options resin infusion

$P_1 > 1$ bar	$P_2 = 1$ bar	resin transfer moulding (RTM) / pressure injection
$P_1 > 1$ bar	$P_2 < 1$ bar	vacuum assisted resin transfer moulding (VARTM)
$P_1 = 1$ bar	$P_2 < 1$ bar	vacuum infusion

AVOID THE LOOSE ENDS EDGE. The new Boeing 7E7 Dreamliner will have bigger windows than current passenger carriers. And not only because composites are stronger than aluminium. No, opportunity knocks because composites have an entirely different way of dealing with stress concentrations than metals. Metals tend to gradually give in through elastic and plastic deformation. They develop cracks that in their turn increase stress concentration until failure occurs. A hole in the skin of an aluminium fuselage with many riveting holes around it is a vulnerable spot. Composites are a different story altogether because their weakness is determined by free edges, places where fibres just end, the same kind of weakness that causes cloth to fray. Therefore sailmakers and tailors sew seams and use metal rings to protect their products.
In composites stress around free edges causes gradual delamination: fibres embedded in polymer with different orientations (anisotropy and heterogeneity) each have their own way of dealing with load in a certain direction, which more often than not differs from the optimum. They wiggle and move about and try to free themselves from the structure. You could say the difference between metals and composites is the difference between materials and structures. The effect is extremely difficult to grasp in terms of lamination theory, but the general rule is that free edges need to be avoided wherever possible. Fibres must never end at places where stress occurs, but always 'lead on' to more relaxed areas in the structure. That is the solution for the windows in a composite fuselage. The whole is not just cut into the skin, but it is cleverly designed, like a groove, in such a way that there is no free edge. In addition the plastic windowpane is not riveted but fitted snugly in the groove from the inside in such a way that it is held firmly in place when the cabin is pressurized.
As usual in composites, some measures to prevent free edge delamination can be borrowed from the world of textiles. Stitching can work, or seaming, or the trick applied to the edges of buttonholes. But the ability to 'freeze' composites in certain three-dimensional shapes provides extra possibilities to prevent loose ends from happening.

Shear coupling mechanism

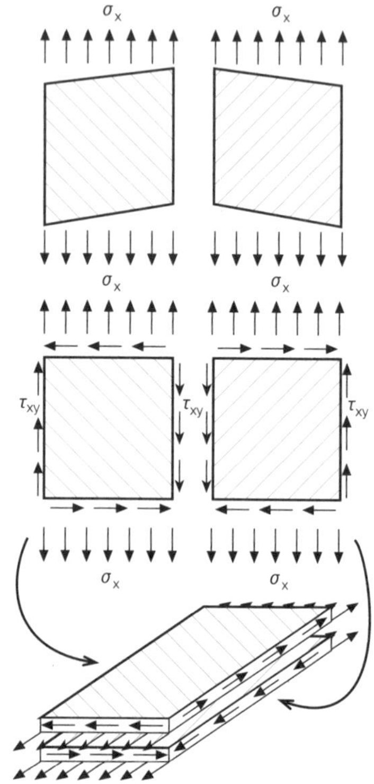

IN DEPTH

Laminate geometry and stresses

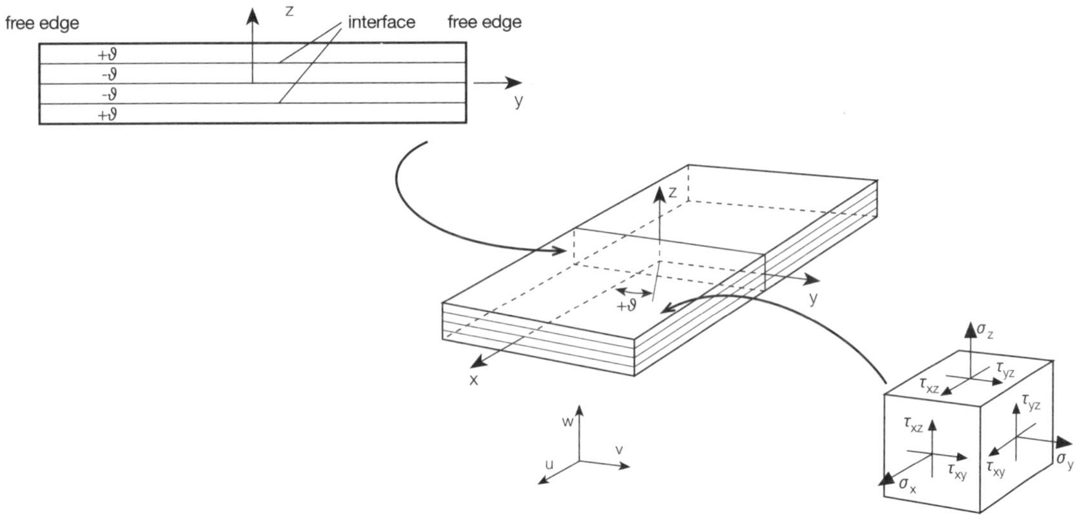

Mechanism of stress transfer in a $(\pm\vartheta)_s$ laminate

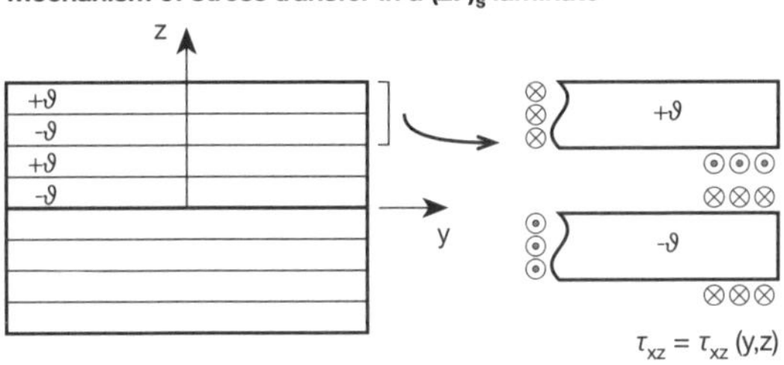

Mechanism of stress transfer in a $(0/90)_s$ laminate

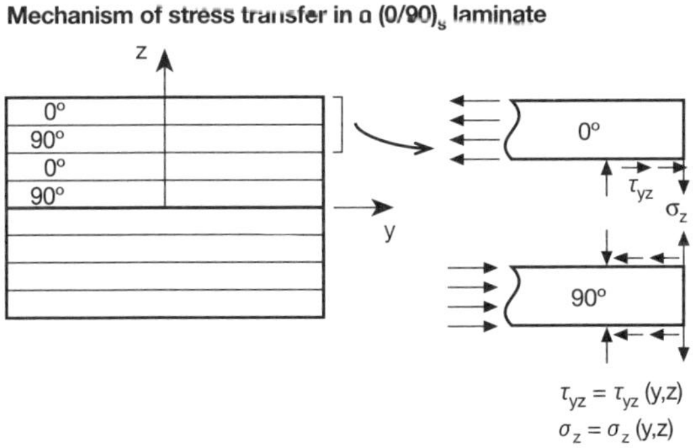

HEALTH PECULIARITIES. Metal and composites behave in very different ways under stress. Around a hole in a piece of aluminium, for instance, tension stress concentration causes a crack that propagates perpendicularly to the direction of the forces. The same conditions in a laminated composite, however, produce a free edge around the hole and will therefore produce a delaminating effect which propagates in the same direction as the forces. So it is not very clever to transplant the blueprint for a metal health monitoring system, with electronic stretch slips and everything, to a composite structure. It won't provide relevant diagnostic information.

Comparison of fatigue cracks in metal and FRP

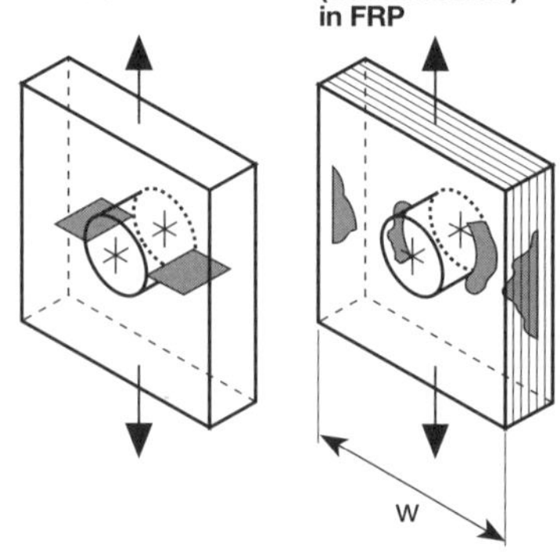

Another reason to beware of black metal is the virtually opposite cracking behaviour in composites.

Comparison of s-n-curves for metal and FRP

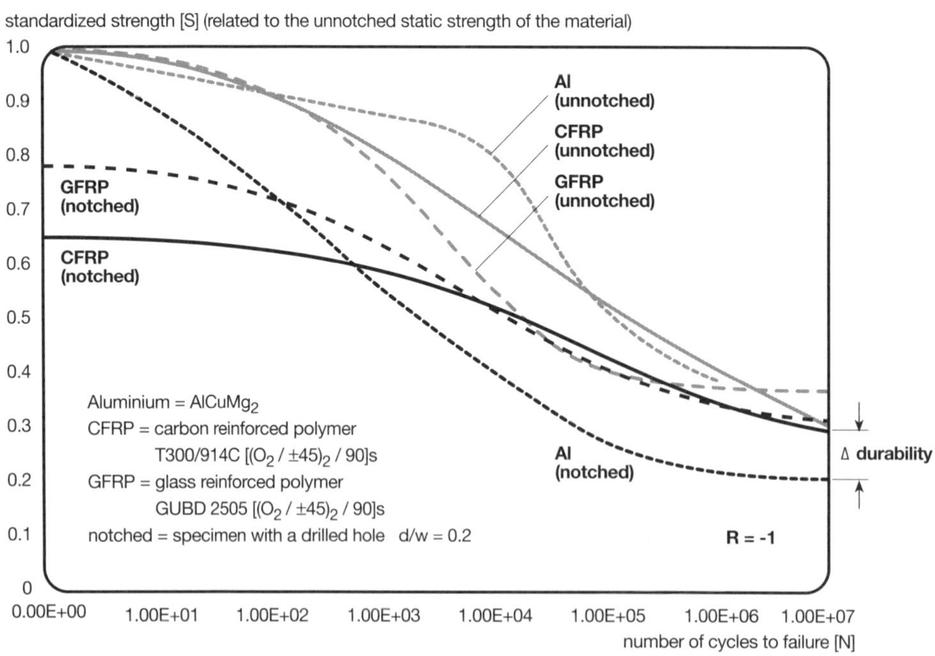

A superb example of metal skin buckling in the wing of a Boeing Stratoliner.

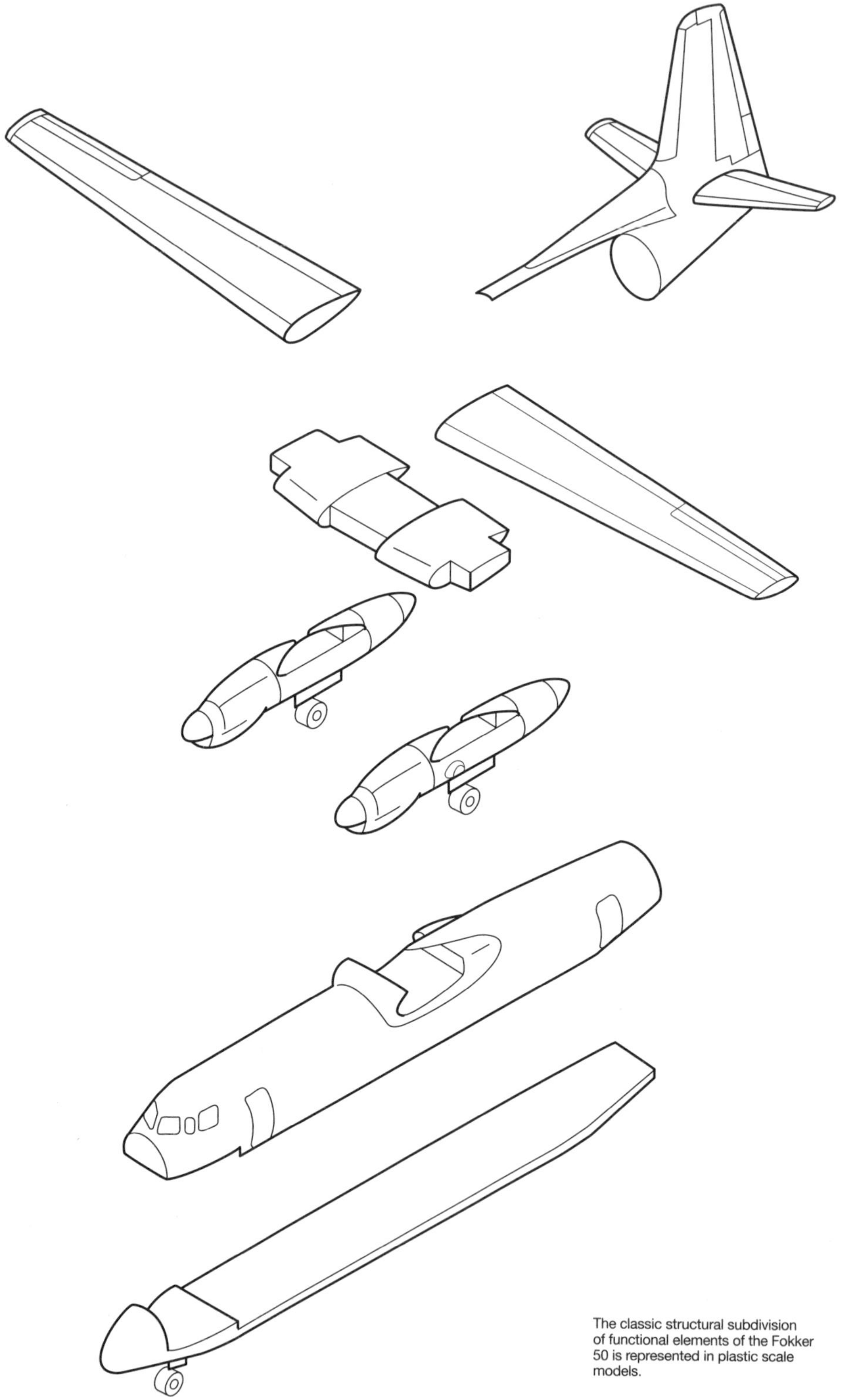

The classic structural subdivision of functional elements of the Fokker 50 is represented in plastic scale models.

Blending wings

Now it is about time we returned to the development of entire lightweight aircraft structures. The flying wing, which could for instance be an 800-seater give or take a 100, provides the perfect opportunity to return to the dream of flying like a (fairly large) bird through the concept of blending wings and fuselage into one integral vehicle. It bears the promise of many of the described structural performance improvements. Its shape and sheer size provide the perfect challenge for most of the structural principles, composite materials and production techniques that we have been discussing so far. Traditionally, and forced by fixed specializations and development organization, aircraft designers do their work from the outside in, sequentially. They start from aerodynamic and economic specifications, and in the resulting shape, usually with a minimum cross-section fuselage, they one by one integrate various functions – insulation, a seating system, control, a pantry – in a more or less standardized structure, in order to achieve subtle and rather predictable improvements. What you get is a house cramped with attempts to provide the illusion of a castle.

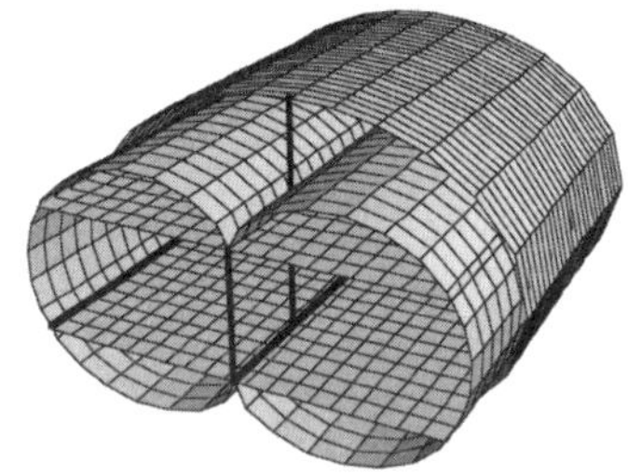

Segregation of pressure cabins and the load carrying structure provides new opportunities for system integration.

There is a hesitant trend in aviation to provide more luxury for passengers. There will be more variation in the price / quality ratio.

That kind of development concept is no longer essential in a voluminous loadbearing fuselage. It is now possible to design from the inside out, in the sense that inside modifications don't have serious repercussions for the overall shape. That provides designers with much more creative freedom.

Moreover separation of functions that traditionally were always combined, such as bearing the load of pressure and serving as a fixture for passenger accessories, can often lead to easier control of all kinds of design issues.
Let us first segregate. Reintegration then will lead to a different concept afterwards. The external upper part of the shell, for instance, only needs to generate and distribute aerodynamic loads. Functionally it has nothing to do with pressurizing a cabin any longer. The same applies to the lower shell of course, but that will need many cutouts for landing gear and other functional parts. Again it does structural work in terms of withstanding forces from the outside, but it has no role in providing passengers with a particular microclimate. For this lower shell we could think of adding energy absorbing elements which could enable controlled, survivable crashes. Think of the composite shock absorbers in Formula 1 cars, but far more sophisticated impact taking is imaginable.

Structural improvement and composite applications have led to very safe Formula 1 racing, which before Jackie Stewart's accident was simply supposed to be dead dangerous.

Most of the systems (climate, cabling, ducts, to name a few) and, with a little extra design effort, the engines and fuel tanks could find a place within this unpressurized shell structure. In the near future tanks could perhaps contain liquid hydrogen. Composites are very well suited to produce safe high pressure gas tanks. The bypass channels of the turbofan engines could then play an important part in noise reduction at the source, which is much more difficult for current external engines, because their fixtures are near-perfect vibration guides. These channels might increase the

SAFELY INTEGRATED SEGREGATION

Its development started at the Laboratory for Lightweight Structures, but now the product is being finished at a new company called ALE (Advanced Lightweight Engineering). It is a composite LPG (Liquefied Petroleum Gas) tank for cars and contrary to what LPG tank experts expected, it is much safer than a steel one. And it is 80 per cent lighter. The tank is wheel-shaped. It fits in the spare wheel compartment. Its skin is a striking example of functional segregation for has an inside layer of impermeable plastic to contain the gas. The outer layer is a simple composite: a inert carbon fibre coiled around the tank to form an isotensoid that is embedded in rubber. The outer layer's function is to withstand the pressure of the enclosed gas.
Most plastics burn, therefore safety rules demand that they are not suitable for application in car LPG tanks. Nevertheless the new tank was allowed to go through a range of tests, the crucial one being to throw the entire tank containing 60 litres of LPG into a bonfire, with everybody at a safe distance holding their hands against their ears. And behold, whereas a steel tank would cause a violent explosion, the plastic one serenely burnt. There was no explosion, because the plastic melted and burnt and let the gas escape quietly. The result demonstrates that rules should not apply to the function of materials and parts but to that of the integrated product. In the meantime several tanks are now travelling around in the spare wheel compartments of Alfa Romeos to prove their value in practice. Development and certification even of 'just' a composite tank takes almost 10 years.

Luigi Colani segregated the parts of a jet engine to arrive at something that looks like the air version of Archimedes' water pump.

air resistance and consequently energy consumption, which we had just diminished considerably through the basic shape of the blended wing concept. Nevertheless this investment in noise reduction could well become paramount in the success of airports and also in the success of aircraft that are allowed to use them. It is one of these difficult environmental dilemmas. But perhaps the exhaust channel could contribute to cleaning the emitted air stream. Charging the exhaust particles electrically and catalysing the gases could lead to miracles of pollution prevention: a modern aircraft designer's dream. In the case of hydrogen fuel that problem of course would not exist. There would have to be some kind of solution for the vapour trail blazing water, however.

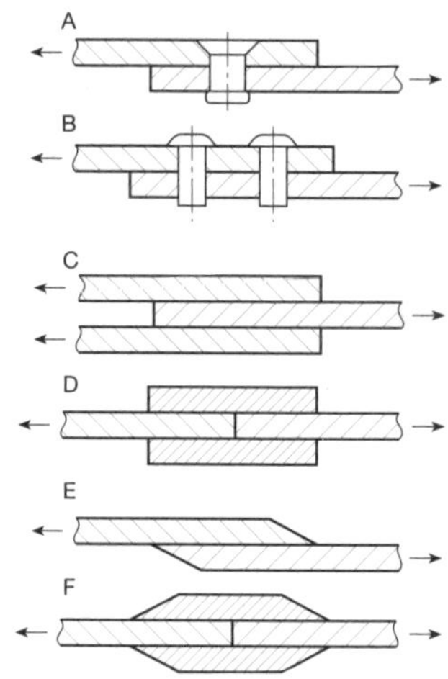

Some standard lap joints to link thin plates with glue or rivets. Lap joints can be riveted with blind rivets (A) or rivets that are accessible from both sides (B). Lap joints have the disadvantage that stress is asymmetrical, leading to bending forces. This also happens in the simple glued lap joint (C). The double lap joint (D) doesn't suffer from this but can be considered inefficient. The strip joint (E) is better and stress concentrations are avoided if the strip has slanted edges (F).

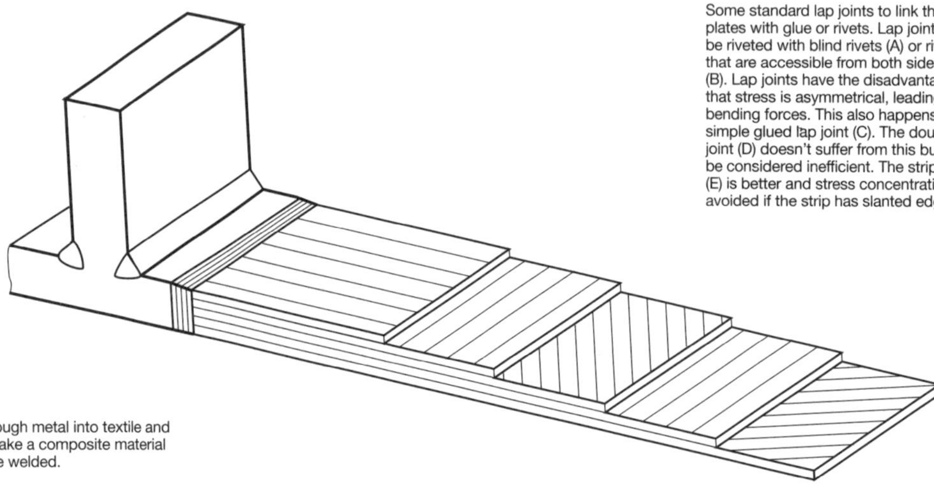

Weave enough metal into textile and you can make a composite material that can be welded.

Back to the external shells. Carbon fibres, or newer, tougher polymer fibres with similar properties, could be combined with metallic fibres and other filling materials in order to obtain the highest possible structural efficiency. A skin like that could withstand mechanical damage, both locally and globally, and it could deal with lightning strikes. Moreover, improved fire safety and safer energy absorption in the event of a crash could already be dealt with at the level of material and element design. As mentioned elsewhere, an interesting idea would be to merge composites with metal fibres

SUCKING THE BOUNDARIES

The more successful aviation got, the less interested airplane designers became in the very principle of flight, of the way in which moving shapes and air interact. Ideally the plane cuts through the air in such a way that this flows around the shape of the airplane in thin laminar boundary layers, but in reality the flow, although generally catenary, is usually more crude and turbulent and complicated. This is partly due to the fact that the airplane shape is not perfect, the other reason being that air does what the weather conditions or maybe another airplane command it to do. So if an aircraft moves through the air and in your thoughts you add all the particles of air to it that are in some way affected by this gross large piece of flying material, what you get is a weird moving cloud of turbulent air that, if we are dealing with a modern wide body, may be at least ten kilometres long. An airplane is a large-scale air manipulator.
However, we can take a different look at the phenomenon of laminar flow to control aircraft movements and air resistance too. The less air is made to move in senseless turbulent ways, the less energy gets wasted. It appears possible to improve aircraft performance by 'handling' air in different ways, such as sucking it in through tiny holes in the surface. The performance of glider airplanes equipped with boundary layer suction is currently being tested. A 500-Watt suction pump, the equivalent of a cheap vacuum cleaner, is sufficient to decrease air resistance by as much as 20 per cent simply by maintaining laminar flow in the boundary layer. Functionally and in terms of scale a glider is scarcely comparable with a large passenger carrier. Nevertheless, in the future suction might be a way to simplify the functions that are now fulfilled by complicated mechanical flaps and rudders and to generally reduce drag. High-speed trains could possibly benefit too if the outer surface were provided with a fine grid of small holes attached to vacuum pumps. Maybe we are on the verge of entering a less turbulent world.

attached to solid metal parts in such a way that composites could be welded or even riveted, or metals parts stitched together, depending on which way you look at it. Because of the design freedom provided by fibres and different kinds of polymers, stability and control once again can become a matter of elastic deformation, as it had been for the Wrights. In combination with chip controlled sensor actuator systems and perhaps advanced boundary layer control, stability during normal flight, without manoeuvring, could become decentralized: the aircraft skin and its pores would 'know' how to behave.

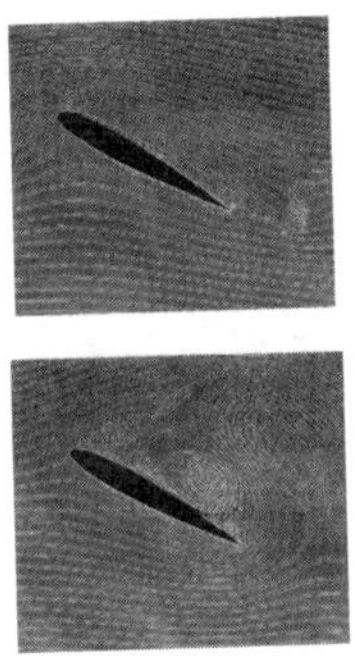

Suction of a boundary layer

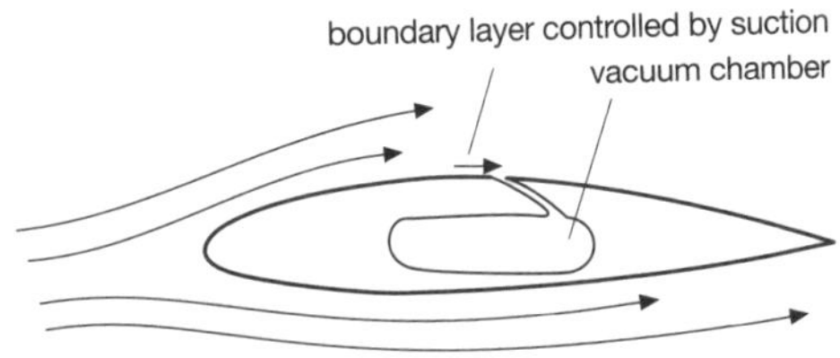

Boundary layers become thinner when exposed to suction from the wing surface, thereby reducing drag.

With regard to the pressure cabin for the passengers and freight, we shall have to switch from the current cylindrical structure to a spatial structure that better fits the available room inside the flying wing. Therefore we will use articulated ultra light pressure vessels that are currently being developed at Delft University. These containers, being flexible, can be folding as well as inflatable. Think of them as huge interlinked soap bubbles. Pressure and the tension in the skins that surround it provide the optimum shapes. They have the advantage that we can deviate from the spherical or certainly not ideal cylindrical fuselage shape if the context demands it, as in the blended wing at hand. These shapes can be made so light because they can be wound or braided. They are isotensoids, for the outlines of possible widenings and constrictions can be easily followed by continuous fibres from pole to pole, without slip or fibre rotation. Stitching more 'cells'

Ceramic plate with integrated metal print labyrinth that heats the bottom in a microwave oven (by Arnout Visser for Rosenthal).

together is possible following the soap bubble example. The design and external loads once again determine the orientation of the reinforcing fibres in the vessel skins. Quick and relatively simple production with specially developed filament winding or braiding machines is imaginable. Pressure vessels like these are perfectly suited to find a place in the flying wing, and become a structural part of it as well, since pressurized vessels can also be considered a large-cell kind of foam, filling the space between the upper and lower shell, thus forming a sandwich panel of a somewhat different scale than is now common.

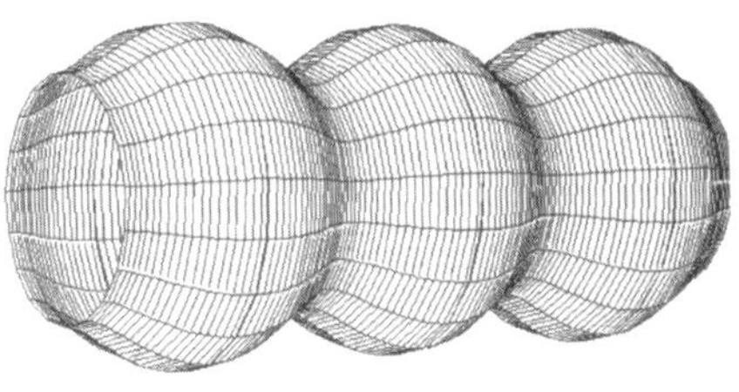

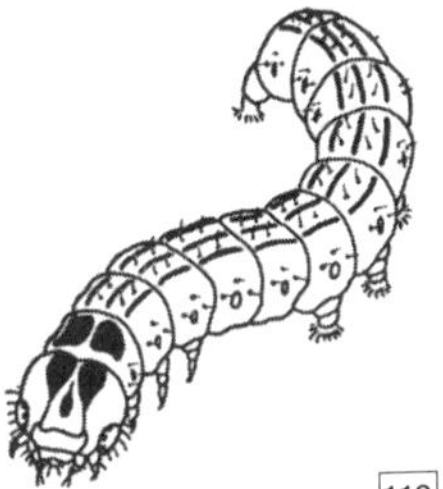

119

An alignment of bubbles can be wound with one fibre. The resulting shape is an isotensoid.

Just like the fibre orientation and matrix consolidation in the gas tanks developed at Delft University, in this structure they can be chosen in such a way that the main load is reduced to constant tension over the entire length of the fibre, the aforementioned isotensoid concept. The topology of this load path strongly resembles a spatial Michell structure. It results in both the lightest and the safest possible pressure vessel. The internal overpressure could even be variable and computer controlled, to ensure that for any external load, the fibres will always be under tension, which is in fact a method of avoiding buckling or instability altogether. Here we have a 'tensegrity' structure that can also serve as a structural element able to sustain both pressure and bending forces. As a singularly articulated structure it can act both as an ultra light pressure cabin and as a pressure vessel for liquified gases.

BACK TO THE CONCEPT

Here we have this famous German industrial designer called Luigi Colani (1928). He calls himself a '3D philosopher', studied aerodynamics when he was young and designed an enormous variety of streamlined industrial products: cars, bottles, wash basins, motorcycles, a typewriter and even a grand piano that looks ready to cruise at 20,000 feet. He also designed aircraft, one of which is a scale model of a huge blended-wingish passenger carrier. Now suppose he were to propose to Boeing or Airbus to develop it into the real thing. What would happen?
'Nothing' is the most likely answer. The design process is not equipped for an entirely new complex product such as this. Aircraft design culture is stuck in habits and in a certain way of dividing up an airplane into functional parts that make it virtually impossible to design something out of the usual. Design departments are extremely good in analysing structures with the aid of computers, so that is precisely what they do. They put a lot of effort into turning the airplane image into sophisticated software models and analyse structural data very thoroughly. But there is no conceptual stage beforehand in which flight characteristics of truly different forms are estimated. Solving the same old structural sums has become more important than the product. Moreover the possibilities of simulating reality in computers has become so true to life that in the end, as in reality, you still don't know what on earth is going on. Our knowledge about the behaviour of air is lagging far behind our understanding of structures that are literally easier to grasp. It would make sense to combine the two in new designs.
There may be a way out of this deadlock. Michel van Tooren is a Professor in Systems Integration Aircraft at the Faculty of Aerospace Engineering of Delft University of Technology. His mission is to integrate the disciplines involved in aircraft design by means of 'Knowledge based Engineering'. The basic principle is that engineering knowledge developed and maintained in a computer supports development from the very start of the process. Van Tooren: 'We are trying to develop geome tric

shapes with behavioural properties that can be estimated. They could serve as primitives, or basic flying elements. Developing an aircraft would then start with combining a number of those building blocks into what could later on be engineered into a structure. This method would allow us to design a broad variety of airplanes with less human effort. So if Colani arrived with rough sketches, they could easily be reduced to knowledge based primitives and manipulated. Instead of "Impossible" we could say "This is interesting. If you would change this and that and add this number of square metres of lift surface, we may have something here.""

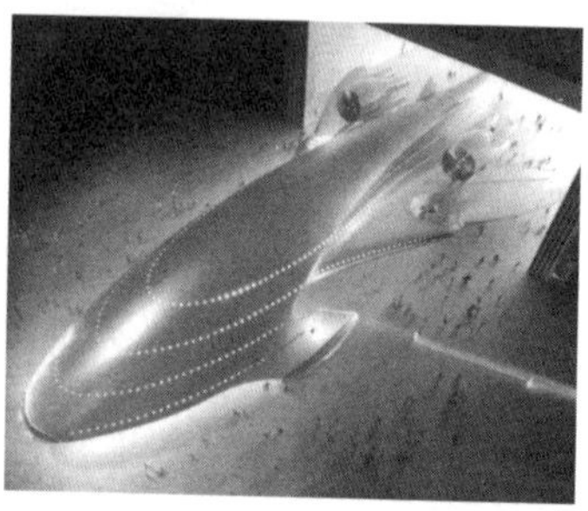

A blended wing body for an 800 passenger aircraft with an average of 2 m^3 living space, with a 13,000 kilometre range at a Mach 0.85 cruise speed, needs about 30 per cent less 'wet' (aerodynamically functional) surface, which implies about the same reduction in terms of dematerialisation. With the same wingspan as a conventional aircraft, the aerodynamic efficiency is improved by about 20 per cent, the result of using a lifting body for payload. Maintaining the same wing profile proportions results in a wing centre that can be as much as four floors high. Three engines are sufficient, and a fuel reduction of no less than 38 per cent is within reach.
The core of the aircraft is now constituted of multiple naturally coupled spaces where passengers can be accommodated. The bubbles are separate from the primary structure and the aircraft systems. The service spaces and corridors can fulfil an important role in the event of an emergency. They can serve as crash- and fire-resistant escape routes. The new concept obsolesces the now common ritual of cabin crew members running up and down long narrow corridors with carts loaded with food and garbage.
From here on we shall not delve any further into the functional and structural advantages and disadvantages of this pressure cabin configuration. After all, there must be something left to do for the young creative engineers that we educate to be observant and to solve design problems. We have shown that with new materials, forms and production processes and above all an open mind, an abundance of product improvements becomes available, especially for aircraft. It remains important to state that the design process that leads to lighter and safer structures, that it should always be firmly founded on the basic and seemingly trivial design rules mentioned and that we should reconsider over and over again.

VESSEL BUILDINGS. Coiled or braided vessels under pressure are light and safe. They could also serve as building blocks for buildings and spatial structures. We can let the non-prismatic articulated 'tank' serve as a model for future three-dimensional spaces that by virtue of overpressure form a stable spatial structure. Here the analogy of blowing bubbles presents itself. Various cells with different diameters can be built together in a stable way. If we combine, for instance, three articulated cylinders we obtain a multidome construction in which the dividing walls, possibly reduced to pillars, become structural elements loaded with pure tension. The fibre orientations within are once again determined by the design and the external loads. When they are large, the separate shell halves produced by means of vacuum injection can be assembled into a multicellular closed space with vertical and horizontal columns or floors and walls.

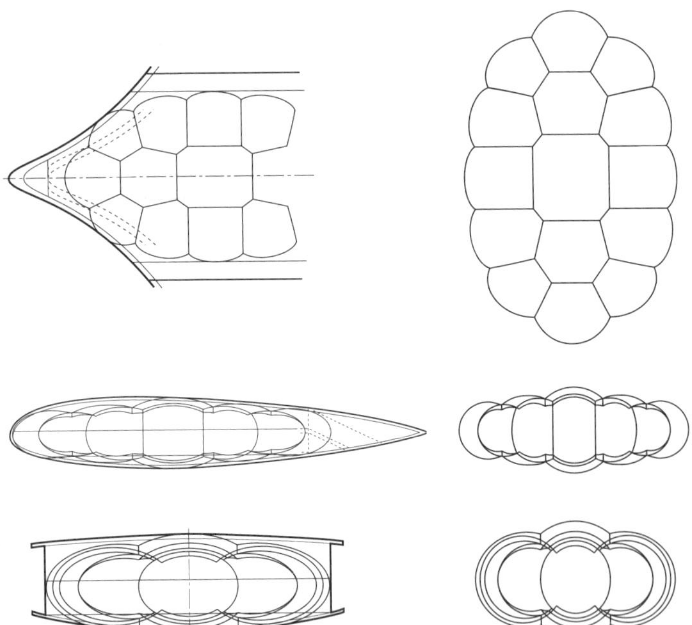

A configuration of 'soap bubbles' would replace the current tube shaped fuselage. It would exclusively have to deal with passenger climate and comfort.

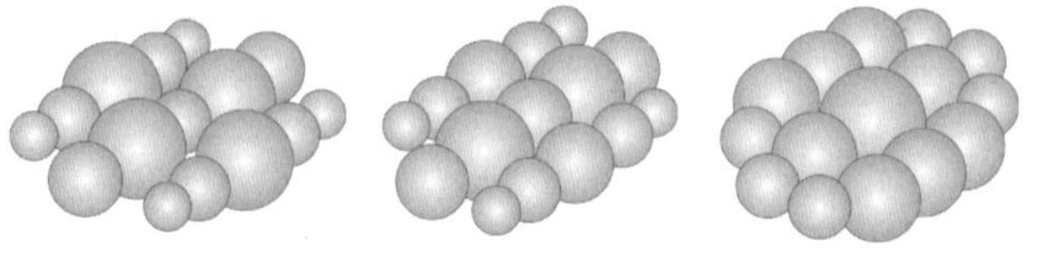

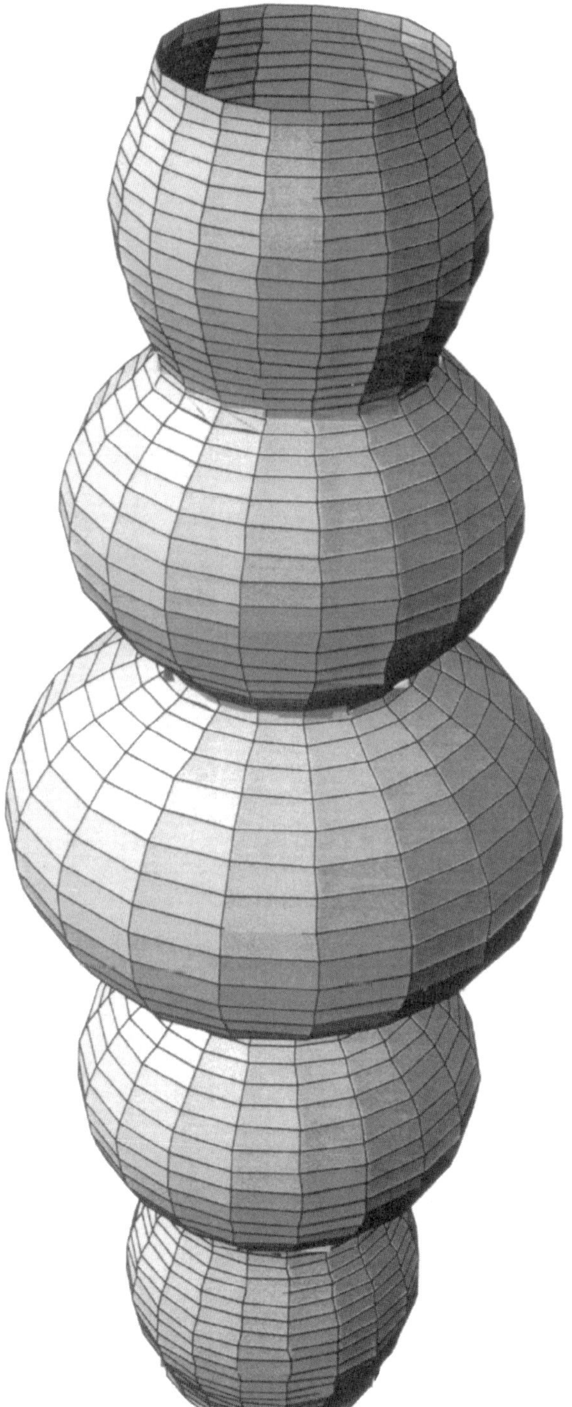

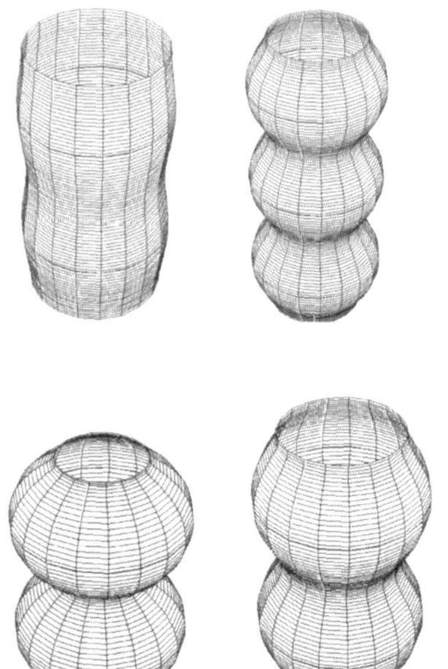

METAL FIBRE COMPOSITES. This may surprise both metal and composite advocates, but it would be most valuable to the designers of the future if, in addition to the broad range of glass, polymer and carbon fibres, they would also have metal fibres at their disposal. For metal producers who are able to upgrade their materials to filaments, a multitude of new application opportunities have arisen thanks to the freedom of product size and shape that composite technology provides. For developers of composite materials and structures, such semi manufactures yield endless new design opportunities. It starts with production itself. Embedded metal fibres can be used for inside-out curing the matrix polymer. There is also the opportunity to create composite parts with metal edges that can be welded together, avoiding the risks of free edges. Structural design, the world of strength, stiffness, deformation behaviour and vibrations, can be extended to a design that includes the desired physical behaviour. This is the world of thermal and mechanical insulation, of conduction and damping; it also enables composites to possess electromagnetic properties. With this technology we can stretch the current purely mechanical design of fibre architecture to include circuits that can control the distribution of energy and information. Equipped with microsensors and actuators this brings us closer to passive and active control of mechanical and physical functions in so-called 'smart structures'. This development takes us far beyond simple metal alloys and the blending of polymers. It implies the creation of new properties and functions by blending and combining all possible fibres. If the envisaged aluminium and steel fibres (table 3) can undergo the same processes as the classical glass and carbon fibres; if the matrix-matrix adhesion of metal fibres and synthetic resins can be adequately tuned; and if, last but not least, the range of matrix materials for metal fibres can be extended to metals that can be sprayed onto the fibres by means of plasma sputter techniques the number of applications becomes endless, from very low to very high temperatures. Using relatively little energy, complex shell structures, from small to very large, can be designed and manufactured with integrated properties that are required in all design sectors.

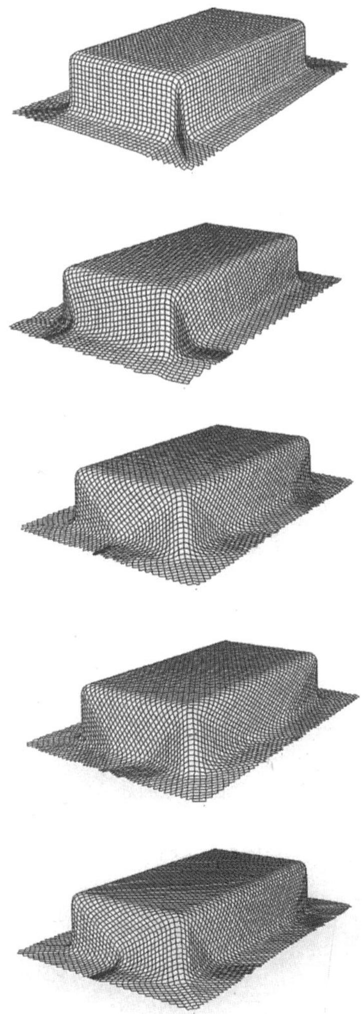

There are limits to textile deformation, and there are opportunities. One can explore the boundaries through draping and pressing.

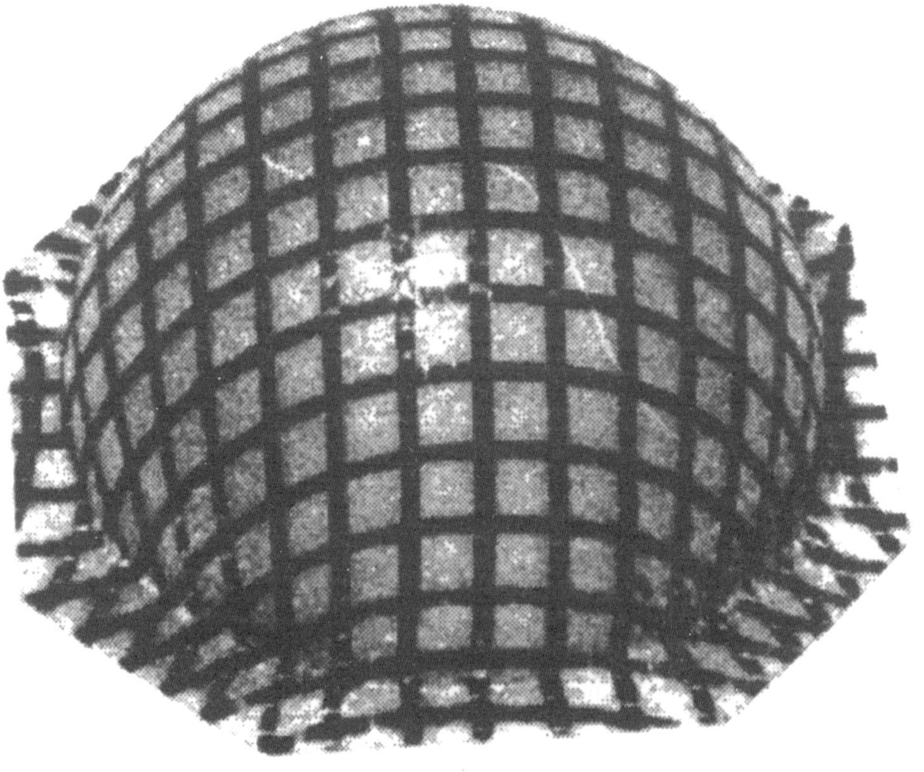

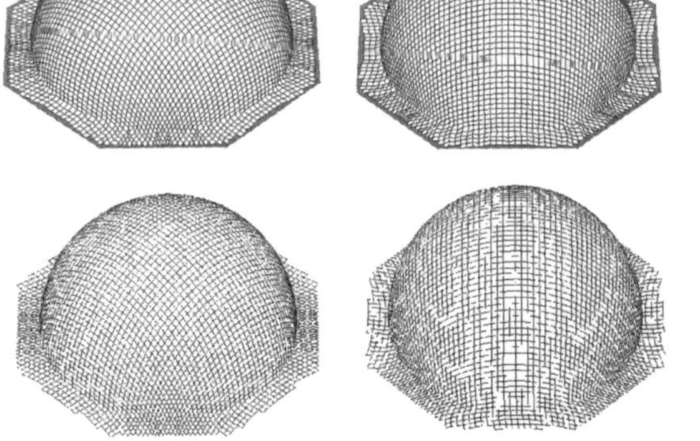

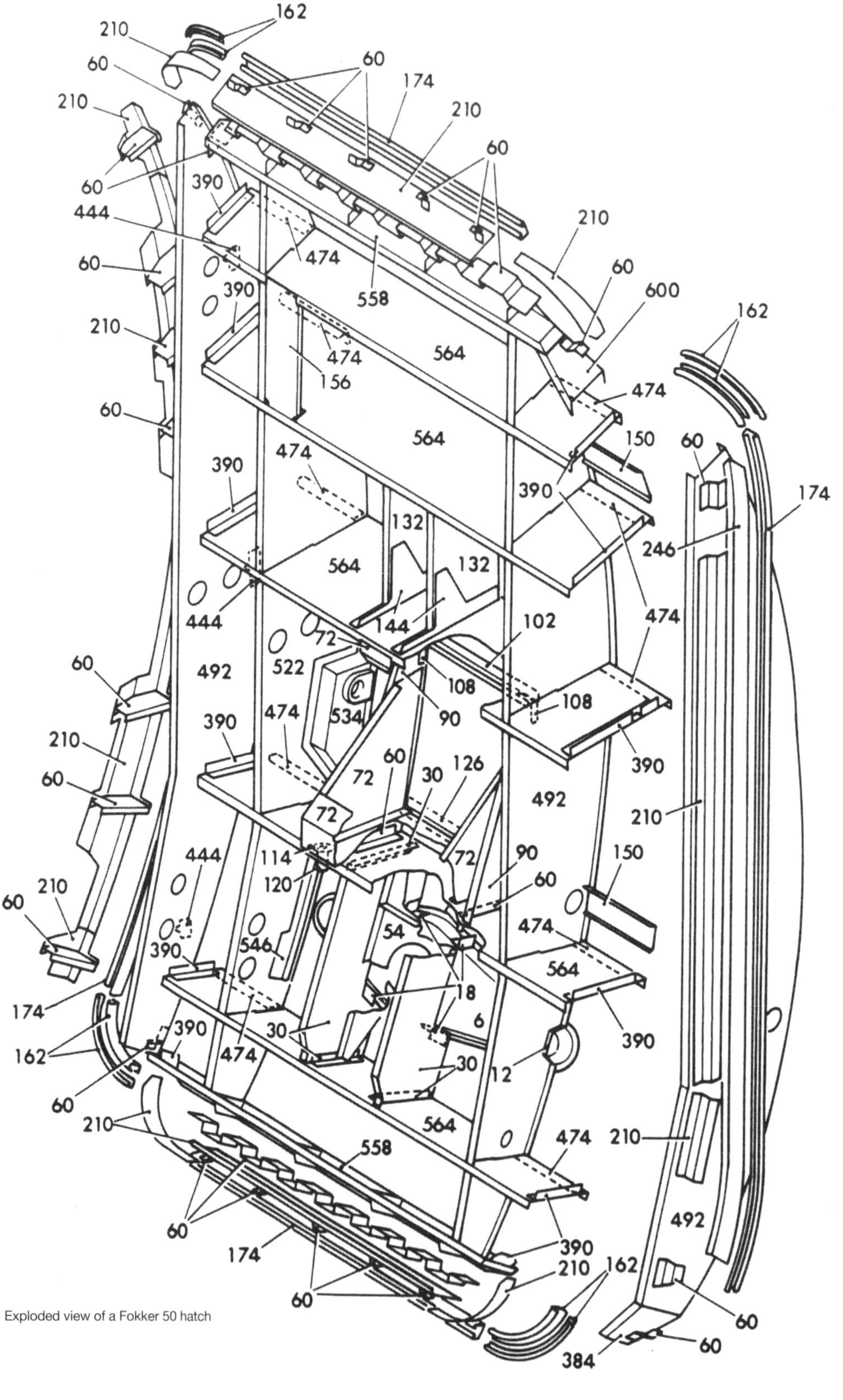

Exploded view of a Fokker 50 hatch

For experts

Some of the information in this book almost drowns within the context in which it is offered. Therefore we have repeated some tables in this section and added some extras on structures and materials for quick reference, especially for those who do not just read books, but apply them as well.

Three transport systems. Specific drag is defined as the amount of resistance that has to be overcome, per unit weight. Fuel efficiency depends on drag, which depends on speed, which is connected to travelling distance and the amount of time people are prepared to spend being on their way from A to B. Therefore different categories of vehicles feature optimal fuel efficiency for different domains of speed. Three systems are likely to survive for local, regional and continental markets respectively: 'smart' road transport, high velocity trains and subsonic aircraft. The word 'smart', incidentally, indicates built-in alertness of vehicles with regard to their position in traffic by means of satellite navigation and/or sensor-actuator systems. It implies a shift from passive to active safety, and considerable weight reduction because of that.

Transport systems, velocity domains and resistance per unit weight

Proven and future transport systems

velocity domain [km/h]	transport market/ system	fuel	dominant resistance / drag	specific drag D_{spec}
$50 < V \quad 0$	LOCAL smart buses, smart cars, smart trains	gas and electricity	wheels (rubber/asphalt) (steel/steel)	$0.01 < D_{spec} < 0.02$
$50 < V < 250$	LOCAL, REGIONAL CONTINENTAL human controlled cars and buses	liquid fuels	wheels and air friction (rubber/stone)	$0.01 < D_{spec} < 0.20$
$125 < V < 300$	REGIONAL, CONTINENTAL high velocity trains	electricity	air friction (steel/steel)	$0.01 < D_{spec} < 0.05$
$300 < V < 900$ relative speed	REGIONAL, CONTINENTAL INTERCONTINENTAL subsonic aircraft	liquid fuels hydrogen	air friction & induced drag	$0.08 < D_{spec} < 0.05$!

Exotic or obsolete transport systems

velocity domain [km/h]	transport market/ system	fuel	dominant resistance / drag	specific drag D_{spec}
$V > 1{,}000$ relative	INTERCONTINENTAL supersonic aircraft	liquid fuels	induced drag and wave drag	$0.10 < D_{spec} < 0.15$
$V < 150$ relative	REGIONAL wings in ground effect	liquid fuels	air friction and induced drag	$0.20 < D_{spec} < 0.30$
$V < 120$ relative	REGIONAL airships	liquid fuels	air friction	$0.025 < D_{spec} < 0.20$

$D_{spec} = 0.1$ means: for 1 kg (mass) in transport about 1 N (force) is needed to overcome resistance

smart could mean: speed and separation control, or satellite navigation control, or active instead of heavy passive safety precautions

Specific properties

Material	Density[A]	Young's modulus[B]	Yield stress[C]	E[D]	E[E]	E[F]	σ[G]	σ[H]	σ[I]
METAL									
steel: 0.2 carbon quenched	7.8	210	773	27	59	0.8	99	108	3.6
steel: piano wires/springs	7.8	210	3100	27	59	0.8	397	273	7.1
aluminium 2024 T3	2.78	72.4	345	26	97	1.5	124	177	6.7
ANIMAL*									
bovine bone	2.1	22.6	-254	11	72	1.3	-121	191	7.6
ivory	1.9	17.5	217	9.2	70	1.4	114	190	7.8
buffalo horn	1.3	2.65	-124	2.0	40	1.1	-95	191	8.6
sinew	1.3	1.24	103	0.95	27	0.8	79	169	7.8
HARDWOOD*									
birch	0.65	16.5	137	25	198	3.9	211	409	18
wych elm	0.55	10.9	105	20	190	4.0	191	405	19
ash	0.69	13.4	165	19	168	3.4	239	436	19
oak	0.96	13	97	14	119	2.4	101	220	10
elm	0.46	7	68	15	182	4.2	148	362	18
SOFTWOOD*									
scots pine	0.46	9.9	89	22	216	4.7	193	433	21
taxus brevifolia	0.63	10	116	16	159	3.4	184	378	17
NATURAL FIBRES* "									
jute	1.46	10-25	400-800	12	91	1.7	411	487	17
flax	1.54	40-85	800-2000	41	162	2.6	909	813	24
sisal	1.33	46	700	35	161	2.7	526	593	20
hemp	1.48	26-30	550-900	19	113	2.1	490	545	18
coir	1.25	6	221	4.8	62	1.5	177	292	12
cotton	1.51	1-12	400-900	4.3	53	1.2	430	497	17
SYNTHETIC FIBRES"									
HM-Carbon (M40)	1.83	392	2740	214	342	4.0	1497	1070	29
HT-Carbon (T300)	1.76	230	3530	131	272	3.5	2006	1317	34
HM-Aramid	1.45	133	3500	92	252	3.5	2414	1590	41
S/R Glass	2.48	88	4590	35	120	1.8	1851	1114	27
E-Glass	2.58	73	3450	28	105	1.6	1337	885	23

* = average values
" = fibres applied in undirectional composites, fibre volume content 50%
grey value properties are only of academic importance

A: ρ [10^3kg/m^3]
B: E [10^9N/m^2]
C: σ [10^9N/m^2]
D: E/ρ [$10^6 \cdot$ m^2/s^2]
E: $E^{1/2}/\rho$ [m$^{5/2}$/s · kg$^{2/3}$]
F: $E^{1/3}/\rho$ [m$^{8/3}$/s$^{2/3}$ · kg$^{2/3}$]
G: σ/ρ [$10^3 \cdot$ m^2/s^2]
H: $\sigma^{2/3}/\rho$ [m$^{7/3}$/s$^{4/3}$ · kg$^{1/3}$]
I: $\sigma^{1/2}/\rho$ [m$^{5/2}$/s · kg$^{1/2}$]

Engineering constants and structure efficiency of different yarns

				carbon fibre yarns	glass fibre yarns	alu fibre yarn	steel fibre yarn
				existing T 300	existing E-type	virtual 7075	virtual piano wire
Material properties	density	ρ	$[10^3\ kg/m^3]$	1.76	2.58	2.70	7.80
	young's modulus	E	$[10^9\ N/m^2]$	230	73	71	210
	yield stress	σ_y	$[10^6\ N/m^2]$	3530	3450	420	3100

Structure efficiency of different yarns

Structure efficiency versus the lowest listed value (1.0)	**based on imaginary yarn properties of equal appearance**	**dominant parameter**	**carbon fibre yarns**	**glass fibre yarns**	**alu fibre yarn**	**steel fibre yarn**
SOLID SHELLS						
Compression	buckling critical	$[E^{1/3}/\rho]$ vs. $[E^{1/3}/\rho]_{min}$	**4.5**	2.1	2.0	**1.0**
CURVED BEAMS	buckling critical or	$[E^{1/2}/\rho]$ vs.	**4.7**	1.8	1.7	**1.0**
Compression	stiffness critical	$[E^{1/2}/\rho]_{min}$	**4.1**	4.1	1.0	3.9
Bending	strength critical					
Bending		$[\sigma_f^{2/3}/\rho]$ vs. $[\sigma_f^{2/3}/\rho]_{min}$				
PRESSURE VESSELS	strength critical, isotensoid wound	$[\sigma_f/\rho]$ vs. $[\sigma_f/\rho]_{min}$	**12.9**	8.6	1.0	2.5
SANDWICH SHELLS						
Bending	stiffness critical	$[E/\rho]$ vs. $[E/\rho]_{min}$	**4.8**	**1.0**	**1.0**	**1.0**

note 1 Properties of natural fibres are very variable, the figures shown are averages in tension and collected from many different publications.

note 2 The yield stress is the lowest stress at which a material undergoes plastic deformation or failure.

note 3 The material values for fibres applied in unidirectional fibre reinforced polymer composites must be adjusted by a proper rule of mixtures.

note 4 A virtual yarn containing a fifty fifty mixture of 7075 T73510 filaments and HM-aramid fibres.

The most important thing to do when choosing a material for a certain function is to keep an open mind. In his book on materials selection Ashby proposes several quick and easy selection procedures. One of them deals with materials and function, regardless of shape. Materials have many properties: density, strength, thermal conductivity, cost, etcetera. The choice of the right combination of properties can be quite difficult. An easy way out is the use of a performance index, a formula to calculate the best solution.
This example from his book of course presents an overview of criteria for minimizing weight. The lightest combination of prop-

Performance-maximizing property groups

mode of loading		**minimize weight for given** stiffness	**minimize weight for given** ductile strength
Tie (tensile strut) F, *l*, specified r free	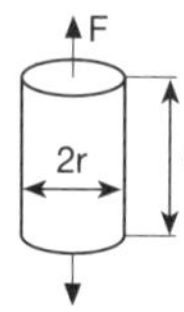	$\frac{E}{\rho}$	$\frac{\sigma_f}{\rho}$
Torsion bar T, *l*, specified r free	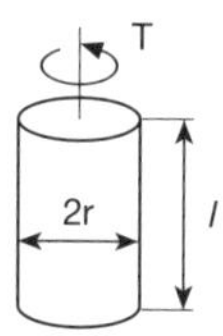	$\frac{G^{1/2}}{\rho}$	$\frac{\sigma_f^{2/3}}{\rho}$
Torsion tube T, *l*, r specified t free	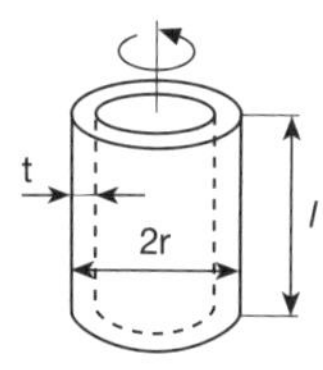	$\frac{E^{1/2}}{\rho}$	$\frac{\sigma_f^{2/3}}{\rho}$
Bending of rods and tubes F, *l*, specified r or t free	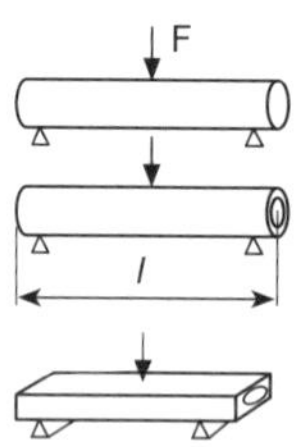	$\frac{E^{1/2}}{\rho}$	$\frac{\sigma_f^{2/3}}{\rho}$
Bucking of slender column or tube F, *l*, specified r or t free	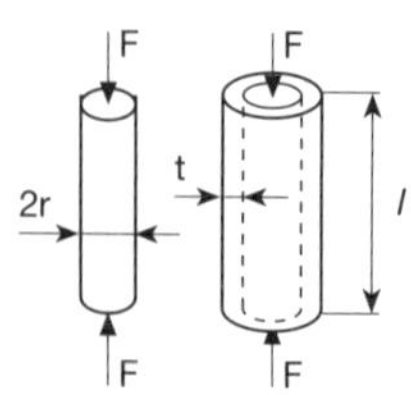	$\frac{E^{1/2}}{\rho}$	-

erties depends on geometry and failure mode. The largest possible calculated value for the formulas represents a minimum weight solution. Two failure modes have been selected: excessive elastic deflection (including buckling) and plastic collapse, for a number of different materials. Some variables are free. In the case of tension (ties) the section can vary. Bending (beams) allows free but proportional choice of width and depth. Plates can have any thickness, and so on. It's all in the table.

Performance-maximizing property groups

mode of loading		minimize weight for given stiffness	 ductile strength
Bending of plate F, *l*, w specified t free	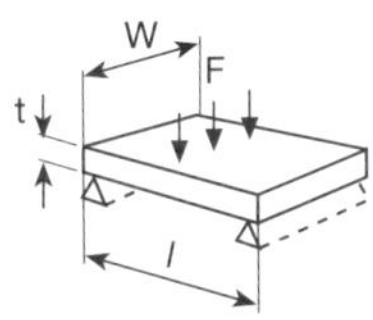	$\frac{E^{1/3}}{\rho}$	$\frac{\sigma_f^{1/2}}{\rho}$
Buckling of plate F, *l*, w specified t free	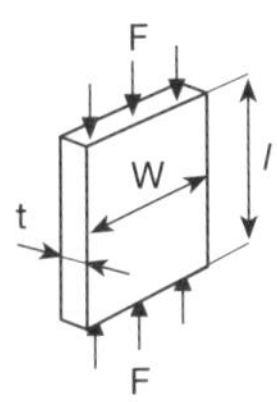	$\frac{E^{1/3}}{\rho}$	-
Cylinder with internal pressure P, r specified t free	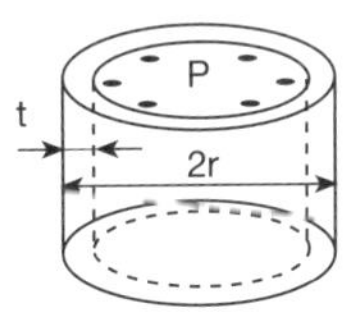	$\frac{E}{\rho}$	$\frac{\sigma_f}{\rho}$
Rotating cylinder w,r specified t free	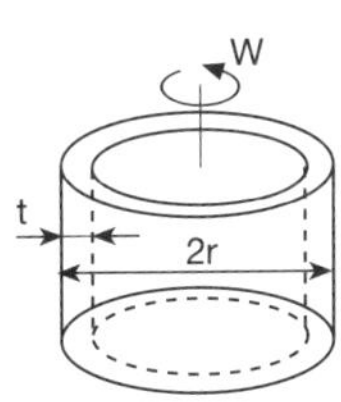	$\frac{E}{\rho}$	$\frac{\sigma_f}{\rho}$
Sphere with internal pressure p, r specified t free	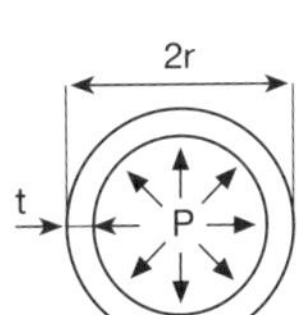	$\frac{E}{(1-\nu)\,\rho}$	-

Ashby's graph tells us that there have been enormous amounts of conceptual jumps within the development of metal structures in a relatively short period of time. The now occuring widening of the polymer and the composite zone shows that starting S-curves with these materials is very promising indeed.

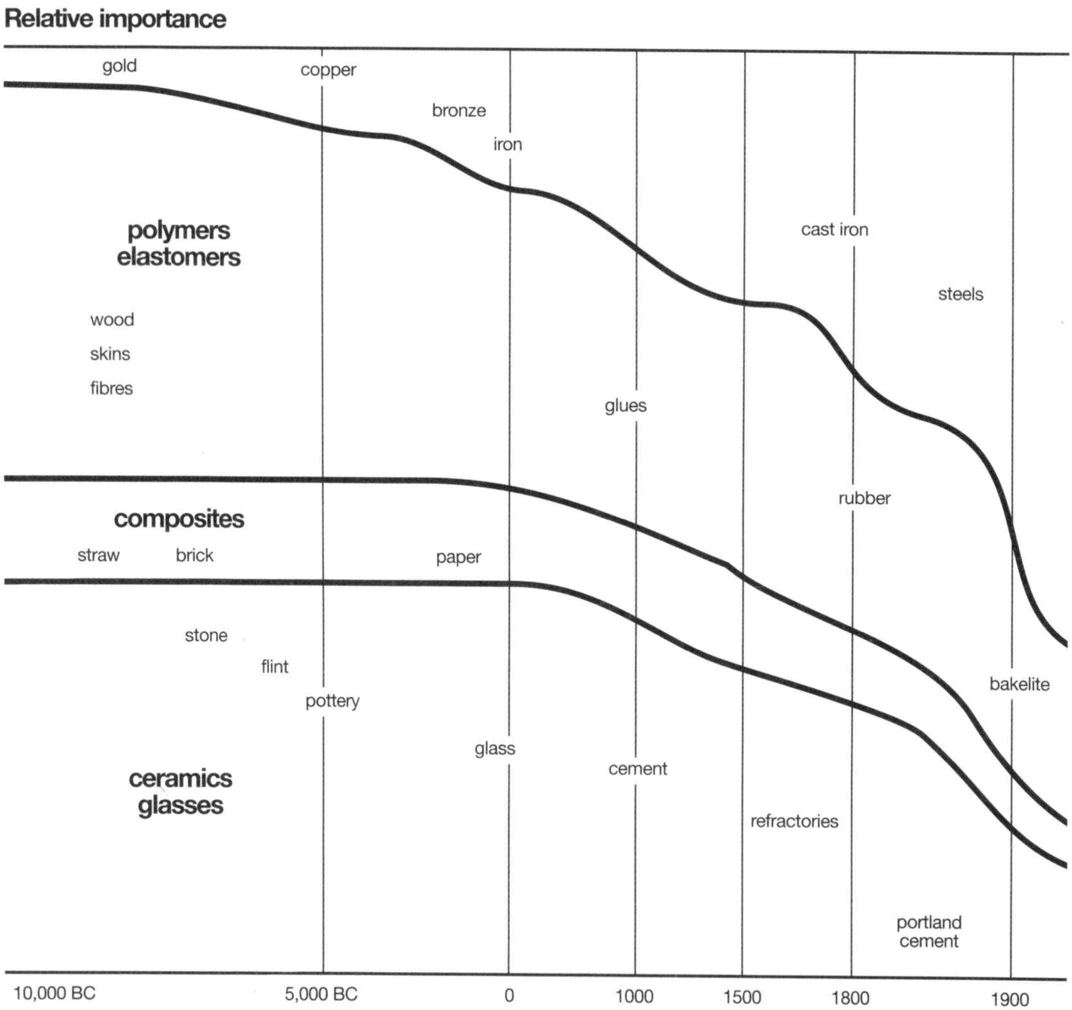

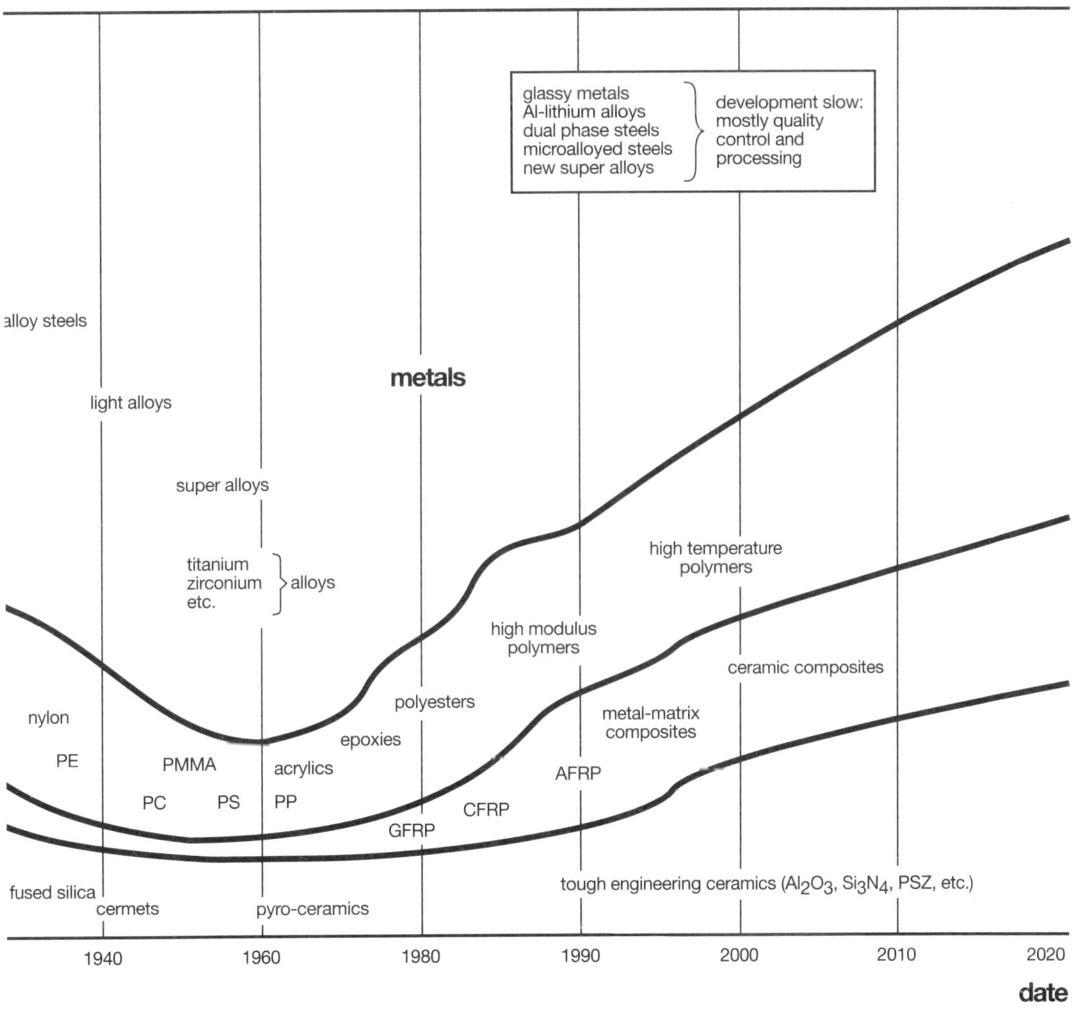
glassy metals
Al-lithium alloys
dual phase steels
microalloyed steels
new super alloys
development slow: mostly quality control and processing
alloy steels
light alloys
metals
super alloys
titanium
zirconium
etc.
alloys
high temperature polymers
high modulus polymers
ceramic composites
nylon
polyesters
metal-matrix composites
epoxies
PE
PMMA
acrylics
AFRP
PC
PS
PP
CFRP
GFRP
tough engineering ceramics (Al_2O_3, Si_3N_4, PSZ, etc.)
fused silica
cermets
pyro-ceramics
1940
1960
1980
1990
2000
2010
2020
date

THERMOPLASTICS

ISO ABBREVIATION	Chemical name Trade names
ABS	**Acrylonitrile-Butadiene-Styrene** Cycolac, Novodur, Ronfalin, Terluran
ARAMID	**Aromatic Polyamide** Kevlar, Twaron (fibre)
ASA	**Acrylate-Styrene-Acrylonitrile** Luran S
CA	**Cellulose-Acetate** Cellidor A, S en U, Cellonex, Ecaron, Setilithe
CAB	**Cellulose-Acetate-Butyrate** Cabulite, Cellidor B
CN	**Cellulose-Nitrate** Celluloid
CY	**Cellulose-Propionate** Cellidor CP
ETFE	**Ethylenetetrafluoroethylene** Tefzel 200, Hostaflon ET
PA	**Polyamide (nylon)** Akulon, Durethan, Ertalon, Grilamid, Grilon, Maranyl, Minlon, Nylatron, Perlon (fibre), Rilsan, Trogamid, Ultramid, Vestamid, Zytel
PB	**Polybutylene** Vestolen BT
PBTP (or PTMT)	**Polybutylene Terephthalate** Crastin, Deroton, Arnite, Kelanex, Orgater, Pibiter, Pocan, Valox
PC	**Polycarbonate** Lexan, Makrolon, Plakrofol N (foil), Pokalon
PEEK	**Polyesteresterketone**
PEHD (or HDPE)	**Polyethylene High Density (hard)** Alkathene, Baylon, Epratene, Hostalen, Lupolen, Marlex, Multilene, Stamylan, Supralen, Vestolen A
PELD (or LDPE)	**Polyethylene Low Density (soft)** Alkathene, Baylon, Fertene, Hostalen LD, Lupolen, Neopolen, Stamylan, Supralen, Trocellen, Vulcathene
PES	**Polyethersulfone** Victrex
PET	**Polyethylene Terephthalate** Arnite, Ertalyte, Hostadur, Rynite, Hostaphan (foil), Melinex, Mylar, Dacron (fibre), Diolen (fibre), Terlenka (fibre), Trevira (fibre)
PF	**Phenolformaldehyde** Aerophenal, Alberit, Alresen, Bakelite, Epratex, Trolitan
PF hapa	**PF laminated with paper** Pertinax, Tufnol
PF hawe	**PF laminated with textile** Celoron, Novotext
PI	**Polyimide** Vespel (resin), Kapton (foil)
PMMA	**Polymethylynemeacrylate** Casocryl, Diakon, Acrylaat, Lucite, Perspex, Polymethacrylester, Plexiglas, Resartglas, Vedril, Vitredil
PO	**Phenoxy resin**
POM	**Polyoxymethylene** Alkon, Celcon, Delrin, Polyacetaal, Ertacetal, Hostaform, Kematal, Ultraform
PP	**Polypropylene** Carlona P, Hostalen PP, Luparen, Moplen, Novolen, Propathene
PPO (modified)	**Polyphenyleneoxide** Ertaphenyl, Noryl
PPS	**Polyphenylenesulfide** Ryton
PS	**Polystyrene** Hostyreen, Styron, Trolitul, Vestyron, Styrodur (foam) and Styrofoam (foam)
PTFE	**Polytetrafluoroethylene** Algoflon, Eriflon, Fluon, Gaflon, Hostaflon TF, Teflon
PTMT (or PBTP)	**Polytetramethylene Terephthalate** Tenite PTMT, Ultradur, PTMT-TSG (foam)
PUR	**Polyurethane** Adiprene, Ertane, Aclathan (elast), Baytec (elast), Contilan (elast), Desmoflex (elast), Desmopan (elast), Dorlastan (fibre), Hostalit Z, Lycra (fibre), Pagulan, Phoenolan, Urepan, Vulkolan, Baydur (foam), Bayfill (foam), Bayflex (foam), Bayfit (foam), Baytherm (foam), Noltopren (foam), Trovipor (foam)
PVC	**Polyvinylchloride** Corvic, Hostalit, Trocal, Vestolit, Vinidur, Vinnol, Vinoflex, Welvic
PVC plasticized	Astraglas, Contiplast, Drakavita, Mipolam, Mipoplast, Novolon, Viplast
PVC rigid	Armodur, Astralit, Astratherm, Carina, Conticell, Darvic, Dekadur, Epradur, Komacel (structural foam), Trosiplast, Varlan
PVDF	**Polyvinylidenefluoride** Dyflor 2000, Kynar, Solef, Steonef
PVF	**Polyvinylfluoride** Tedlar
SAN	**Styrene Acrylonitrile** Luran, Lustran, Tyril, Vestoran
SB	**Butadiene-Styrene** Vestyron
VF	**Vulkanfiber** Dynos, Hornex, Zellglas, Cellophan, Cuprophan

THERMOSETS

ISO ABBREVIATION	Chemical name Trade names
EP	**Epoxid resins** Araldit, Bakelite, Epikote, Epotuf, Epoxyn, Lekutherm, Levepox, Melopas, Suraplast
MF	**Melamine-Formaldehyde** Albamit, Bakelite, Formica, Hornitex, Keramin, Melbrite, Melolam, Melopas, Mepal,Resopal, Resart, Supraplast, Ultrapas
PSU	**Polysulfone** Bakelite
UF	**Urea-Formaldehyde** Cibanoid, Gabrite, Kaurit, Mouldrite, Philite, Pollopas, Resopal
UP	**Unsaturated Polyester** Alpolit, Durodet, Keripreg, Leguval, Norsopreg, Novolite, Palatal, Vestopal

ADVANCED COMPOSITES. Composite materials with structural properties comparable to or better than those of aluminium, e.g. boron, carbon and aramid composites.

ALUMINIUM (AL). Metallic element, melting point 660 degrees Celsius, density 2700 kg.m-3. Uses: ubiquitous aerospace alloy base.

ANISOTROPIC. Having mechanical and/or physical properties, which vary in direction relative to natural reference axes in the material.

ARAMID. Type of highly oriented aromatic polymer material. Used primarily as a high-strength fibre.

AXIAL. According to the direction of the axis.

BIDIRECTIONAL LAMINATE. A reinforced plastic laminate with the fibres oriented in two directions in the plane of the laminate.

BINDER. The agent applied to glass mat or preforms to bond the fibres prior to laminating or moulding.

BOND STRENGTH. The amount of adhesion between bonded surfaces; a measure of the stress required to separate a layer of material from the base to which it is bonded.

BRAIDING. A mechanical weaving of fibres into a structural form such as a tube, a square box or a U-shaped channel.

BREAKOUT. Separation or breakage of fibres when the edges of a composite are drilled or cut.

BREATHER. Loosely woven material that does not come into contact with the resin but serves as a continuous vacuum path over a part in production.

BRIDGING. Fabric plies over a curved edge that do not come into full contact with the core material. Also excess resin that has formed on edges during curing.

BUCKLING. Unstable displacement of a structural part, such as a panel, caused by excessive compression and/or shear. Microbuckling of fibres in a composite material can also occur under axial compression.

BUNDLE. General term for a collection of essentially parallel filaments.

BULK MOULDING COMPOUND (BMC). A premixed blend of thermosetting resin, reinforcements, catalysts and fillers for use in compression-, transfer- or injection moulding processes.

CARBON (C). Non-metallic element, melting point 3550 degrees Celsius, density 2200 kg.m-3. Uses: carbon reinforcement fibres.

CARBON FIBRE. Fibre produced by the pyrolysis of organic precursor fibres, such as rayon, polyacrylonitrile, in an inert environment. The term is often used interchangeably with the term graphite; carbon fibres and graphite fibres do, however, differ. The basic differences lie in the temperature at which the fibres are made and heat-treated, and in the amount of elemental carbon produced.

CATENARY. Uniformity of strand length in a specified length of roving stretched under tension. Poor catenary means some strands in the roving length are longer than others.

CENTRIFUGAL CASTING. A processing technique for fabricating cylindrical structures, in which the composite material is positioned inside a hollow mandrel designed to be heated and rotated as the resin is cured.

CERAMIC. Chemical compound or mixture of compounds with high temperature capabilities, often metal (or silicon) oxides, nitrides or carbides.

CERAMIC MATRIX COMPOSITE. Composite material in which the matrix is a ceramic.

CFRP. Carbon fibre reinforced plastic.

CFRTP. Carbon fibre reinforced thermoplastic.

CIRCUMFERENTIAL WINDING. Process of winding fibre perpendicular to the axis.

COMPOSITE (MATERIAL). Three-dimensional combination of at least two materials differing in form or composition, with a distinct interface separating the components.

CONTINUOUS FILAMENT YARN. Yarn formed by twisting two or more continuous filaments into a single continuous strand.

CONTINUOUS REINFORCEMENT. Where reinforcement fibres or filaments pass uninterrupted through the composite.

CORE. In sandwich construction, the central component to which inner and outer skins are attached.

CRACK. An actual separation of moulding material visible on opposite surfaces of the part and extending through the thickness.

CRAZING. Region of ultrafine cracks that may develop on or under resin surface.

CREEP. The change in dimensions of a material, under constant load over a period of time.

CURE. To change the properties of a resin by chemical reaction, which may be condensation or addition, usually by the action of heat or catalyst, or both, and with or without pressure.

CURVATURE. Geometric measure of the bending and twisting of a plate.

DAMAGE TOLERANCE. The ability of a structure to carry loads in the presence of flaws or defects or other damage for a specified period of use without repair.

DAMPING. Diminishing the intensity of vibrations.

DEFORMATION. Changes in size and shape of a body resulting from externally applied stresses, temperature change and moisture absorption.

DELAMINATION. Physical separation or loss of bond of the layers of materials in a laminate.

DIELECTRIC STRENGTH. The voltage required to penetrate insulating material.

DIMENSIONAL STABILITY. Zero or nearly zero deformation due to change in temperature and moisture content. A property of carbon/resin and aramid/resin composite materials that is exploited in the design of high-accuracy components.

DOUBLER. An extra layer of reinforcement, applied to increase stiffness or strength in portions of a laminate expected to incur abrupt load changes. May be counterproductive in certain cases of bending load.

DRAPE. The ability of a fabric or prepreg to conform to a contoured surface.

DRY WINDING. In filament winding, when fibre reinforcement is applied to the mandrel without first being wetted out with resin.

E-GLASS. Abbreviation for 'electrical glass', borosilicate glass fibres, which have high electrical resistance. Most often used in conventional polymer matrix composites.

ELASTICITY. The property of materials to immediately recover their original size and shape when load is removed after deformation.

ELASTOMER. Elastic substance occurring naturally as rubber, or produced synthetically as a specific gradation of many polymer materials.

ELONGATION. The fractional increase in length of a material loaded in tension.

ELEMENT. A generic part of a more complex structural member.

END. A strand of roving consisting of a given number of filaments gathered together.

ENGINEERING CONSTANTS. Measured directly from uniaxial tensile and compressive, and pure shear tests applied to unidirectional as well as laminated composites. Typical constants are the effective Young's modulus, Poisson's ratio and shear modulus.

EPOXY RESINS. Thermosetting resins made by polymerization of epoxies with other materials such as amines, alcohols, phenols and unsaturated compounds.

FABRIC. Planar, woven materials constructed by interlacing yarns, fibres or filaments.

FATIGUE. In materials or structures, the cumulative irreversible damage incurred by cyclic mechanical and/or thermal load in given environments. Fatigue can initiate and extend cracks.

FIBRE. Single filament, rolled, formed in one direction and used as the principal constituent of woven and nonwoven composite materials. The most common fibres are glass, boron, carbon and aramid.

FIBRE SURFACE TREATMENT. Carbon fibres are usually given a postgraphitization surface treatment by either a chemical or an electro-mechanical method. The surface treatment enhances the bond between the fibres and the resin matrix in the cured laminate. This improves the shear and compressive properties of laminates.

FILAMENT. A continuous discrete fibre. The cross-section is not necessarily circular.

FILAMENT WINDING. Automated process of placing filaments on a mandrel in a prescribed pattern. Resin impregnation can be before or during the winding, known as prepreg or wet winding. The mandrel can be removed after curing the composite materials. Filament winding is most advantageous in building pressure vessels.

FINITE ELEMENT ANALYSIS. Process of selecting the optimum combination of materials in a structure, based on computer-based computational modelling.

FULL SMART SYSTEM. A system of sensors, interrogation and analysis software and compensators able to monitor and react in a required manner to external influence without human intervention.

GEODESIC. The shortest distance between two points on a curved surface.

GEODESIC ISOTENSOID. Constant-stress level in any given filament at all points in its path.

GEODESIC ISOTENSOID CONTOUR. In filament wound reinforced plastic pressure vessels, a dome contour in which the filaments are placed on geodesic paths so that under pressure loading the filaments will exhibit uniform tension throughout their length.

GFRP. Glass fibre reinforced plastic.

HAND LAY-UP. The process of placing (and working) successive plies of reinforcing materials or resin-impregnated reinforcement in position on a mould by hand.

HARDENER. A substance that may be added to a resin to promote and control the curing process by participating in the curing reaction.

HELICAL. Describes ply laid onto a rotating mandrel at an, often 45°, angle.

HOMOGENEITY. Material uniformity within a body. In the mechanics of composite materials, micro- and macroscopic homogeneity are achieved by smearing the actual heterogeneity.

HONEYCOMB. Manufactured product of resin-impregnated sheet material (paper, glass fabric, etc.) or sheet metal formed into hexagonal-shaped cells; used as a core material in sandwich constructions.

HYBRID COMPOSITE. Composite containing at least two distinct types of matrix or reinforcement.

HYGROSCOPIC. Tending to absorb moisture from the air.

IMPACT DAMAGE. A structural anomaly created by the impact of a foreign object.

IMPREGNATE. In reinforced plastics, to saturate the reinforcement with a resin.

INCLUSION. Physical and mechanical discontinuity occurring within material or part.

IN SITU. In the original position; in filament winding designates a mandrel that remains in place after winding, as opposed to one that is removed.

INJECTION MOULDING. A method of forming a plastic to the desired shape by forcibly injecting heated polymer into a mould.

INTERFACE. The plane formed when two material surfaces make contact. In glass fibres, for instance, the area at which the glasses and sizing meet, and in a laminate the area at which the reinforcement and resin meet.

INTERLAMINAR SHEAR. Shearing force that produces displacement between two layers along the plane of their interface.

IRON (FE). Metallic element, melting point 1538 degrees Celsius, density 7870 kg.m-3. Uses: base element for steels, iron-based alloys, alloying additions.

ISOTROPY. Property that is not directionally dependent (having the same physical and/or mechanical properties in all material directions). Metals are often assumed to be isotropic. This is normally not the case, but they do generally show considerably more isotropy than fibre reinforced composites.

JOINT. The location at which two different parts are held together. It is the general area of contact for a bonded structure. There are several types of bonded joints:
Butt Joint. The edge faces of the two parts are at right angles to their other faces.
Scarf Joint. A joint made by cutting away similar angular segments of two parts and bonding them with the cut areas fitted together.
Lap Joint. A joint made by placing one element partly over another and bonding together the overlapped portions.

KNITTING. Three-dimensional filament knotting process that leads to a loose and flexible fabric.

LAMINATE. Plate consisting of layers of uni- or multidirectional plies of one or more composite materials.

LAP JOINT. A joint made by overlapping two parts and bonding them together.

LAY-UP. To place or the process of placing layers of reinforcing material into position in or on a mould.

LINER. The continuous, usually flexible, reinforced resin barrier on the inside surface of a plastic or thermoset laminate, used to protect it from chemical attack or to prevent leakage under stress.

MANDREL. A form, fixture or male mould used as the base for production of a part in processes, such as lay-up or filament winding.

MATRIX. Material in which reinforcing fibre of a composite is embedded. Matrix materials include thermosetting and thermoplastic polymers, metals and ceramic compounds.

MICROCRACKING. Microscopic cracks formed in composites when thermal stresses locally exceed the strength of the matrix.

MILLED FIBRE. Continuous strands hammer-milled into very short fibres.

MODULUS. Elastic constant defined as the ratio between the applied stresses and the related deformation, such as Young's modulus, shear modulus or stiffness moduli in general.

MONOMER. A single molecule that reacts with like or unlike molecules to form a polymer.

MOULD. The cavity or matrix in or on which the plastic, metal or ceramic composition is placed and from which it takes its form.

MULTIFILAMENT YARN. A multitude of fine, continuous filaments (often 5 to 100), usually with some twist in them to facilitate handling.

PEEL STRENGTH. Bond strength, obtained by peeling the layer.

PITCH. The distance between structural parts such as frames.

PLASTIC. A material that contains as an essential ingredient an organic substance of high molecular weight, is solid in its finished state, and at some stage in its processing into finished articles can be shaped by flow.

POLYMER. A high-molecular-weight organic compound, natural or synthetic, whose structure can be represented by a repeated small unit, e.g. polyethylene. Synthetic polymers are formed by addition or condensation polymerization of monomers. Some polymers are elastomers, some plastics.

POROSITY. Voids remaining in a material or component after manufacture, e.g. gas bubbles in fibre reinforced resin products, gas or vacuum voids in metal castings. The porosity of a material is normally expressed as the ratio of the volume of air or void contained within the boundaries of a material to the total volume.

PREFORM. Pre-shaped fibrous reinforcement, supplied without matrix, but often containing a binder to maintain shape.

PREPREG. Fibrous reinforcement (sheet, tape, tow, fabric or mat) pre-impregnated with resin and capable of storage for later use.

PRIMARY STRUCTURE. An aerospace critical loadbearing structure.

PULTRUSION. Continuous process for manufacturing rods, tubes, profiles and other linear structures that have a constant cross-section. The process involves drawing continuous reinforcements through a resin-impregnation bath, and then pulling the wetout material through a heated shaping die, where cure takes place.

PYROLYSIS. The subjection of an organic compound to very high temperatures.

REGRIND. Scrap composites reground into pellets or powder for reuse in moulding new parts, either as a new base material or in combination with virgin materials.

REINFORCED REACTION INJECTION MOULDING (RRIM). A closed moulding process that mixes two highly reactive resin components for cure. Reinforcement is added to one of the resin components.

REINFORCED PLASTIC. A plastic with strength properties greatly superior to those of the base resin.

REINFORCEMENT. A strong inert material bonded into a plastic, metal or ceramic to improve its strength, stiffness and impact resistance. Reinforcements are usually long fibres in woven or non-woven form. To be effective, the reinforcing material must form a strong adhesive bond with the matrix.

RESIN. A solid, semi-solid or pseudo-solid organic material, which has an indefinite (often high) molecular weight, exhibits a tendency to flow when subjected to stress and usually has a softening or melting range. Most resins are polymers.

RESIN TRANSFER MOULDING (RTM). A closed moulding process, in which catalysed resin is transferred into an enclosed mould cavity to impregnate a pre-positioned fibrous reinforcement.

ROVING. Loose assemblage of filaments with no twist. Roving can be impregnated for use in filament winding, braiding and in unidirectional tapes.

S-GLASS. The standard abbreviation for 'structural glass'. This is a magnesia/alumina/silicate glass fibre reinforcement designed to provide the tensile strength required in high-performance composites.

SAFETY FACTOR. The ratio of the ultimate design loads to the limit of applied loads.

SANDWICH STRUCTURE. A composite component featuring a lightweight core material (usually honeycomb, foam or balsa wood) placed between two relatively thin, dense, functional skins.

SECONDARY STRUCTURE. Aerospace structure that is not critical to flight safety.

SHEAR. An action or stress resulting from force applied in a direction parallel to the plane of adhesion between the surfaces of two adjacent components or layers, causing or tending to cause one to slide relative to the other.

SPRAY-UP. Technique with a spray gun as the processing tool. In reinforced plastics, for example, glass and resin can be simultaneously deposited in a mould. In essence, roving is fed through a chopper and ejected into a resin stream which is then directed at the mould.

STAPLE. Collection of short filaments of a spinnable length.

STIFFNESS. Measure of the resistance of a material to deformation. Young's modulus designates the stiffness of a material subjected to uniaxial stress; shear modulus to shear stress. For composite materials, stiffness and other properties are dependent on the orientation of the material.

STRAND. A primary bundle of continuous filaments combined in a single compact unit.

STRENGTH. The maximum stress a material can sustain. Like the stiffness of a composite material, this is highly dependent on orientation.

STRESS. Intensity of forces within a body. The normal components induce length or volume change and the shear component, shape change.

STRESS CONCENTRATION.
Micromechanical level
Concentration occurs at the fibre-matrix interface.
Macromechanical level
Concentration occurs at holes, notches, termination points, joints, etc.

SYNTHETIC FIBRE. Fibre made of materials other than glass or carbon, such as polyester or aramid.

TAPE. Thin unidirectional prepreg.

TENSILE STRENGTH. The maximum tensile stress sustained by a test specimen before failure.

THERMAL CONDUCTIVITY. The ability to conduct heat.

THERMOPLASTICS. Plastics that can repeatedly become soft when heated, without change of properties.

THERMOSET. Polymer that sets in a chemical curing reaction and cannot be returned to the uncured state.

TITANIUM. Metallic element, melting point 1670°C, density 4540 kg.m-3. Matrix alloy for composites, structural material for aerospace use, generally where operating temperatures exceed those possible with aluminium.

TOUGHNESS. The ability of a material to absorb energy.

TOW. Continuous, ordered assembly of essentially parallel, collimated filaments.

ULTRAVIOLET CURE. The process of curing resins and adhesives with ultraviolet light.

UNIDIRECTIONAL (UD). General term denoting orientation of fibres in one direction.

VACUUM-BAG MOULDING. Moulding technique wherein a part laid up on an open mould is cured under a layer of sealed film from which entrapped air has been removed by vacuum.

VEIL. An ultra thin, nonwoven mat, often composed of organic fibres as well as glass fibres and used primarily as a corrosion barrier.

VISCOSITY. The property of resistance to flow exhibited within the body of a material, measured in comparison with water. The higher the number, the greater the resistance to flow.

WARP. The yarn running lengthwise, parallel to the length of the loom, in a woven fabric.

WARP(AGE). A change in dimension of a cured laminate from its initial mould shape.

WATER ABSORPTION. Ratio of the weight of water absorbed by material to the weight of dry material.

WEATHERING. The exposure of materials to the outdoor environment.

WEAVE. To interlace fibres in a pattern, often based on a 0°/90° grid; the pattern formed by interlacing yarns. Interlacing patterns vary. In plain weave, for instance, warp and weft fibres alternate to make both fabric faces identical, and in satin weave a pattern is produced by a warp tow over several weft tows and under one weft tow.

WEFT. The fibre bundles in a woven fabric that run transverse to the warp yarns. Also known as fill, or woof.

WET LAY-UP. Application of a resin to dry reinforcements of a mould.

WINDING. Any process in which continuous fibres are applied under controlled tension to a rotating form (mandrel) in a predetermined geometrical relationship to make a structure.

YARN. A continuous, ordered assembly of essentially parallel, collimated filaments with a twist.

YIELD STRENGTH. The stress at which a material exhibits a specified deviation from the proportionality of stress to strain; the beginning of plastic deformation.

YOUNG'S MODULUS. Ratio of normal stress to the corresponding strain for tensile or compressive stresses less than a material's proportional limit (elastic deformation).

ZERO BLEED. Laminate fabrication procedure that does not allow loss of resin during cure.

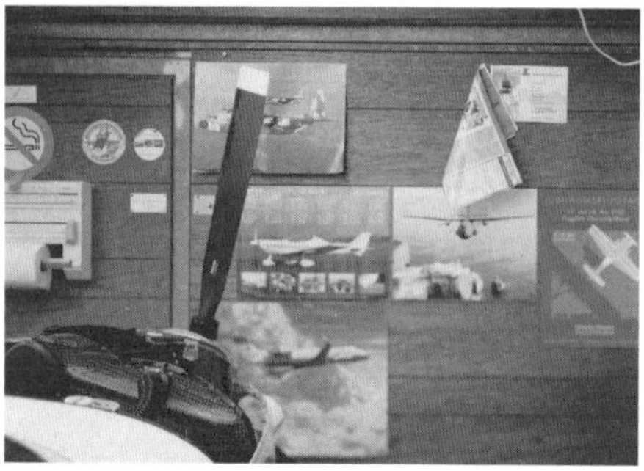

Aad Kievit, Oss

Henk van Aalstede, Amsterdam

Cor Dijkman Dulkes, Nagele

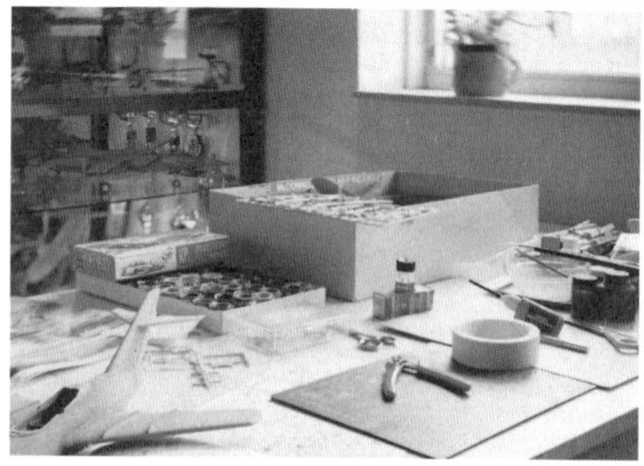

Meindert de Vreeze, Amsterdam

D.A.C. Ranger: Jan Laurier, Evert-Jan Cornet, Den Haag

Stichting Vroege Vogels, Lelystad

Klaas van der Wielen, Utrecht

REFERENCES

Abate, R., Aeroplani Caproni (Italian), Museo Caproni, Trento 1992
Condit, P.M., (Boeing Company), Airline costs on the rise while yield slips, AIAA World Aviation Conference, Los Angeles,1996
Dabrowski, H., Flying Wings of the Horten Brothers, Schiffer Publishing Ltd, Atglen PA, 1995
Dulken, S. van, Patenten op de 19e eeuw, Veen Magazines, Diemen 2003
Dulken, S. van, Patenten op de 20e eeuw, Veen Magazines, Diemen 2003
Gibson, L.J., Ashby, M. F., Cellular Solids, Structure and Properties, Pergamon Press, Oxford 1988
Gibbs-Smith, C.H., A Brief History of Flying: From Myth to Space Travel, Her Majesty's Stationery Office, London 1967
Gordon, J.E., Structures; or why things don't fall down, Penguin Books Ltd., London 1991
Hillier, J., Japanese Prints from 1700 to 1900, Phaidon, Oxford 1985
Jansen, H.A., Schönfeld C.W., Motoren voor de voortstuwing van vliegtuigen (Dutch), Argus, Amsterdam 1960
Kampschuur, B., Vliegtuigmotoren (Dutch), Nederlandsch Uitgeversbedrijf van Wetenschappelijke Uitgaven N.V., Leiden 1944
Keegan, J., A history of warfare, Pimlico, London 1993
Koussios, S., Filament winding; a unified approach, Delft University Press, Delft 2004
Koussios, S., Bergsma, O.K., Beukers, A., Pressurizable structures comprising different surface sections, PCT/NL02/00534, Delft 2002
Krakers, L.A., Tooren, M.J.L. van, Beukers, A., Bergsma, O.K., Multidisciplinary fuselage design: Integration of mechanics & acoustics, proceedings of SAMPE Europe conference, Paris, 2000, pp. 617-622.
Liebeck, R.H., Page, M.A., Rawdon, B.K., Blended-Wing-Body Subsonic Commercial Transport, AAIA-98-0438, January 1998.
Pape, G.R., Campbell, J.M., Northrop Flying Wings, Schiffer Publishing Ltd., Atglen PA, 1994
Pearce, P., Structure in Nature Is a Strategy for Design, The MIT Press, Cambridge MA 02142, 1990
Oneal, M., Greising, D., Boeing best big on a plastic plane, Chicago Tribune, Chicago January 12, 2005
Rothwell, A., Introduction to aircraft structural design, Faculty of Aerospace Engineering, Delft University of Technology, Delft 1992
Schafer, A., Victor, D., The past and future of global mobility, Scientific American, October 1997, pp. 36-39.
Steenderen jr, C. van, Zij maakten luchtvaartgeschiendenis (Dutch), A.W.Bruna & Zoon's uitgevers-maatschappij, Utrecht 1942
Tomlow, J., IL34, Das modell, The model, El modelo, (3 languages), Karl Krämer Verlag, Stuttgart 1989
Torenbeek E., emeritus lecture: 'We nemen een schoon vel tekenpapier', (Dutch), Delft University Press, Delft 2000
Uytenbogaart, J.W.H., Kooy, J.M.J., Ballistics of the Future - V-weapons, Stam, Haarlem 1946
Verpoest, I., Texcomp-2 (New textiles for Composites), Katholieke Universiteit Leuven 1994
Wohl, R., A Passion for Wings; Aviation and the Western Imagination 1908-1918, Yale University Press, New Haven 1994
Westerop, W.C.J. van, Fokker en de twintigste eeuw, Foundation Fokker Heritage Trust / Foundation National Aviation Museum Aviodome-Schiphol, 1998.
Zee, S. van der, Salto Mortale: Fokker in bedrijf 1911-1996 (Dutch), Basalt & D'ARTS, Amsterdam 1998
ANC-23 Bulletin, Sandwich Construction for Aircraft, part 1, Washington D.C. February 1951
The Jet Engine, Rolls-Royce, Derby 1996
VM119, Vezelversterkte kunststoffen in constructieve toepassingen (Dutch), Federatie NRK en Vereniging FME-CWM November 1999

CREDITS

'Flying Lightness' is based on Adriaan Beukers' inaugural lecture at Delft, University of Technology and is meant as an extension of 'Lightness; the inevitable renaissance of minimum energy structures' (010publishers, Rotterdam / ISBN 90-6450-344-6).

The authors wish to thank the Laboratory of Structures and Materials of the Faculty of Aerospace Engineering of Delft University of Technology and the Centre of Lightweight Structures. Personal thanks go to Sotiris Koussios, Michel van Tooren, Otto Bergsma, Lester Lardenoye, NIMR, Doshisma University, Kyoto and all others who were, knowingly or unknowingly, involved in the making of this book.

Graphic design Erik Wong and Eva Heisterkamp
Graphs Talea Bohlander
Superfly photography / cover Ingmar Swalue
Copy editor John Kirkpatrick
Printed by Snoeck-Ducaju & Zoon, Ghent
General assistance Lisette Vollmer

Photography: Reuters: Concorde nose cone (p. 13), ANP: Tall Alexander Sizonenko (p. 48), AP: Freighter cross-section (p. 52-53), AP: Broken windmill (p.89), Romi Chiorean (www.romiphoto.com): Airplane cemetery Mojave Desert (p.39), Greet de Schipper: Truck drivers (p.16), Roel Visser: Traffic jam (p. 8), Tim Dirven: KLM dance, (p. 40-41), Marcel Portenge: Antarctic vehicle, (p. 97)

ISBN 90 6450 538 1